AGRICULTURAL BIOTECHNOLOGY

AGRICULTURAL BIOTECHNOLOGY

By

Ashok Kumar

Lecturer

Department of Zoology

Bundelkhand University

Campus Department

Jhansi

DISCOVERY PUBLISHING HOUSE

NEW DELHI-110002

First Published – 2006

Reprinted – 2017

ISBN: 978-81-8356-117-4

Agricultural Biotechnology

Published by:

DISCOVERY PUBLISHING HOUSE PVT. LTD.
4383/4B, Ansari Road Darya Ganj
New Delhi - 110 002 (India)
Phone: +91-11-23279245, 43596064-65
Fax: +91-11-23253475
E-mail: discoverypublishinghouse@gmail.com
sales@discoverypublishinggroup.com
web: www.discoverypublishinggroup.com

Printed at:
Infinity Imaging Systems
Delhi

Preface

Agricultural Biotechnology in diverse facets is presently advancing in meteoric speed. By the time the present title could be completed agriculture biotechnology may have advanced further in exciting direction. The very broad stretch of agriculture biotechnology has never been brought within two covers. A student being exposed for the first time to this interesting subject needs to be given an idea about it. The present title has tried to do this precisely. All kinds of pollutants affecting the agricultural crops directly or indirectly via soil, water or air have been touched to create an impression about the diversity. The enormous prospect of application of biotechnology in profitably using agricultural wastes and products have been indicated. On the whole graduating students both in agriculture and science may find interest in this book will benefit having one at hand.

There can be no claim to originality except in the manner of treatment and much of the information has been obtained from the books and scientific journals available in different libraries.

The author expresses his thanks to his friends and colleagues whose constant inspiration have initiated him to bring out this book.

The author expresses his gratitute to Mr. Wasan and staff of M/s Discovery Publishing House for their whole hearted co-operation in the publication of this book.

Author

CONTENTS

1 Introduction

Technological advances in biotechnology, including genetic engineering, have enabled transfer of genetic traits both within species and between entirely different plant and animal species. Currently, biotechnology techniques are being used in various fields, including agriculture, veterinary medicine, pharmaceutical development, forestry, energy conservation, and waste treatment. These techniques, if applied responsibly, have the potential to increase productivity in crops and livestock, control pests, produce new food and fiber crops, and develop effective medicines,. Potential environmental and forestry through improved nutrient availability in crops and livestock, use of fewer artificial inputs (e.g., synthetic nitrogen fertilizers, insecticides, and fungicides), and more cost effective and environmentally friendly waste management practices, such as bioremediation.

If realized, these improvements will help protect ecological systems by reducing habitat degradation. In addition, some of the biotechnology techniques should improve the economics of agricultural and forestry production aystems. Although genetic engineering can be expected to provide major benefits to agriculture and the environment, risks with the use of this technology should also be recognized. In this chapter, we assess the environmental, health, and socioeconomic benefits and risks of biotechnology, including genetic engineering, in agricultural systems.

DISEASE-RESISTANT CROPS

Resistance against crop disease in plants, caused by viruses, bacteria and fungi is now being explored through biotechnology and genetic engineering techniques as a way to reduce the loss of crops. Because viruses in the field cannot easily be treated, the production of genetically engineered, virus-resistant crops is agronomically significant. In addition, few antibacterial chemicals are available to control bacterial diseases. It has been estimated that viruses, bacteria, and fungi are collectively responsible for significant crop losses estimated at 12%, or nine hundred million tons, of preharvest yield worldwide.

More than 350 field tests of genetically engineered disease-resistant plants have been approved in the United States since 1987, and the majority of these have been created to produce disease-resistant, genetically-engineered crops impervious to viral infections. Success in engineering virus resistance in tobacco, alfalfa, potato, cucumber (*Cucumis sativus*), melon (*Cucumis melo*), alfalfa, and tomato plants have been reported by respectively. (See Table 1.1.)

Field trials with tobacco containing the gene from the mosaic virus for the production of the coat protein have shown that resistance can be transgenically induced. For example, in China, field trials of tobacco that contains the tobacco mosaic virus and tomatoes with cucumber mosaic virus are under way. Efforts are also being aimed at rice because of its importance as a staple crop in this region. In Japan, a method for producing fertile transgenic rice plants using an electroporation system has been developed. Transgenic rice plants expressing the rice stripe virus–coat protein (RSV–CP) have been developed to fight the rice stripe virus, one of the major viruses of rice plants in Japan, Korea, China, and Taiwan.

The findings of a 3-year biosafety study of ecological risks have demonstrated that expressing the introduced gene (RSV–CP) in a japonica rice variety (Kinuhikari) resulted in transgenic rice plants that did not: (1) affect morphological and ecological traits with the exception of some somaclonal variations, (2) hybridize with closely grown rice plants, (3) exhibit the tendency to become weeds, (4) produce any detectable toxic substances, and (5) have any observable effects on subsequent cultivation, microorgansims in soil insects in florae, or on surrounding plants. However, these results will need to be followed up with longer-term biosafety assessment in the future.

TABLE 1.1 : PLANTS GENETICALLY ENGINEERED FOR VIRUS RESISTANCE THAT HAVE BEEN APPROVED FOR FIELD TESTS

Crop	Disease(s)	Research organization
Alfalfa	Alfalfa mosaic virus, Tobacco mosaic virus (TMV), Cucumber mosaic virus (CMV)	Pioneer Hi-Bred
Barley	Barley yellow dwaft virus (BYDV)	USDA
Beets	Beet necrotic yellow vein virus	Betaseed
Cantelope and squash	CMV, papaya ringspot virus (PRV) Zucchini yellow mosaic virus (ZYMV), Waltermelon mosaic virus II (WMVII)	Upjohn
	CMV	Harris Morna Seed
	ZYMV	Michigan State University
	ZYMV	Rogers NK Seed
	Soybean mosaic virus (SMV)	Cornell University
	SMV, CMV	New York State Experiment Station
Corn	Maize dwarf mosaic virus (MDMV) Maize chlorotic mottle virus (MCMV, Maize chlorotic dwarf virus (MCDV)	Pioneer Hi-Bred
	MDMV	Northup King
	MDMV	EdKelb
	MDMV	Rogers NK Seed
Cucumbers	CMV	New York State Experiment Station
Lettuce	Tomato spotted wilt virus (TSWV)	Upjohn
Papayas	PRV	University of Howaii
Peanuts	TSWV	Agracetus
Plum Trees	PRV, plum pox virus	USDA
Potatoes	Potato leaf roll virus (PLRV), Potato virus X (PVX), Potato virus Y (PVY)	Monsanto
	PLRV, PVY, late blight of potatoes	Frito-Lay

Table 1.1 Contd.

Crop	Disease(s)	Research organization
Potatoes	PLRV	Calgene
	PLRV, PRY	University of Idaho
	PLRV, PVY	North Carolina State University
Soybeans	SMV	Pioneer Hi-Bred
Tobacco	ALMV, tobacco etch virus (TEV), Tobacco vein motting virus	
	TEV, PVY	University of Florida
	TEV, PVY	North Carolina State University
	TMV	Oklahoma State University
	TEV	USDA
Tomatoes	TMV	Monsanto
	CMV, tomato yellow leafcurl virus	
	TMV,ToMV	Upjohn
	ToMV	Rogers NK Seed
	CMV	PetoSeed
	CMV	Asgrow
	CMV	Harris Moran Seeds
	CMV	New York State Experiment Station
	CMV	USDA

In the United States, squash and the papaya are two of the more recent models of crop engineering for virus resistance. In 1994, the genetically engineered,virus-resistant squash developed by Asgrow seed company for resistance to zucchini yellow mosaic virus (ZYMV) and watermelon mosaic virus II (WMV II) was one of the first genetically engineered crops commercialized in the United States. Researchers have also developed two genetically engineered papaya lines by utilizing rDNA techniques to isolate and clone a papaya ringspot virus (PRV) that encodes for the production of the viral coat protein. Papaya, one of the three largest crops in Hawaii, has been decimated in recent years by PVR. Hawaiian papaya growers believe these two lines of genetically engineered, disease resistant papaya could save the $45 million Hawaiian papaya industry from extinction. However, it should be noted that, although the mild strain of PRV displayed excellent resistance to PRV isolates from Hawaii, it showed only moderate to no protection to isolates from different geographic regions (e.g.Bahamas, Mexico, China, Brazil and Australia).

ENGINEERING BACTERIAL AND FUNGI RESISTANCE IN CROPS

Harms (1992), who has reviewed recent developments in the production of resistance to fungal and bacterial diseases via genetic engineering points out that, although there has been much research on how to incorporate such resistance into crop plants, few improvements have been made in this area. Chen and Gu (1993) have described efforts that are being made to combat bacterial blight, which can reduce rice yields by as much as 10%, through genetic engineering.

Developing disease-resistant crops should also receive high priority secondary to the large anounts of fungicides that are currently applied to fruit and vegetable crops. Aspelin et al. (1994) reported that in 1993 131 million pounds of pesticidal active ingredient was applied at a cost of $584 million. Fungicides are sometimes harmful to beneficial insects and toxic to earthworms and many other beneficial soil biota. The number and activity of these soil biota are important in maintaining soil fertility over time because they recycle nutrients in organic matter and aid in water percolation and soil aeration. Furthermore, fungicides rank highest for carcinogenicity of all pesticides applied to agriculture and account for approximately 70% of human health problems associated with pesticide use.

One way to reduce crop losses to fungi and the external application of fungicides is to introduce genes that encode proteins with antifungal properties into crop plants. Several genes have been identified so far in fungi, bacteria, and plants that are effective for the engineering of resistance to fungi based on their ability to produce enzymes, such as chitinase, that attack the cell wall of fungi. Transgenic tobacco plants with a chitinase gene from beans produced elevated levels of chitinase in roots and leaves compared with control plants in greenhouse experiments. Both experimental and control plants were grown in soil inoculated with the fungal pathogen *Rhizoctonia solani*. A positive association was also found between the level of chitinase expressed in the experimental plants and survival. Broglie et al. (1993) and Lin et al. (1995) also reported some success has been achieved with engineering resistance to the stem pathogen (*Rhizoctonia solani*) in oilseed rape or canola (*Brassica napus*) and rice, respectively. The engineering of resistance to the fugnus *Fusarium oxysporum* in the tomato has also been successful.

Creation of New Weeds

In terms of risks, it has been proposed that large-scale cultivation of plants expressing viral and bacterial genes could lead to novel ecological risks. The most significant ecological risk would be gene transfer via pollination from cultivated crops to wild relatives. For example, it has been postulated that the virus-resistant squash (*Cucurbita pepo*), which is native to the southern United States, where it is an agricultural weed. If the virus-resistance genes were to spread, newly disease-resistant wild squash could become a hardier, more abundant weed. Moreover, because the United States is the origin for squash, changes in the genetic make-up of wild squash could lessen its value to squash breeders.

Another area of concern is the production of virus-resistant sugar beets, which is likely to result in exchange of genes between cultivated and wild populations of beets (*Beta vulgaris L*) because production areas contain wild or weed beet populations, or both, separated by only a few kilometers. A genetic exchange could take place owing to wind pollination, biotic pollination, or the common gynoniecy of wild beets. A genetic introgression from seed beet to wild beet popluations has already been observed in Europe.

Viruses that Infect New Hosts

Some plant pathologists also hypothesize that development of virus-resistant crops may allow viruses to infect new hosts through transencapsidation. This may be especially important for certain viruses (e.g., luteoviruses) for which possible heterologus encapsidation other viral RNAs with the expressed coat protein is known to occur naturally. With other viruses, such as the PRV, risk of heteroencapsidation is thought to be minial because the papaya itself is infected by very few viruses.

Creation of New Viruses

Virus-resistant crops may also lead to the creation of new viruses through an exchange of genetic material or recombination between RNa virus genomes. Recombination between RNA virus genomes requires infection of the same host cell with two or more viruses. Several authors have pointed out that recombination may also occur in genetically engineered plants expressing viral sequences on infection with a single virus and that large-scale cultivation of such plants may lead to increased possibilities of combinations. It has recently been shown that an RNA transcribed from a transgene can recombine with an infecting virus to produce highly virulent new viruses.

An overall strategy of risk assessment utilizing an incremental approach entails: (1) identifying potential hazards, (2) determining frequency of recombination between homologous but nonidentical sequences, and (3) determining whether such recombinants can have already demonstrated that, even though a particular pseudorecombinant (resulting from pseudorecombination or the situation in which gene components of one virus are exchanged with the proteins of another coat) had enhanced fitness relative to either of the other original strains.

HERBICIDE-RESISTANT CROPS (HRCs)

At the moment at least 4 engineered crops for target herbicide resistance are on the market, and 13 among the key crops in world production have been extensively tested in field trials (Table 1.2). In addition, some crops (e.g., corn) are being engineered to contain both herbicide (Glyphosate) and insecticide-resistance (BT δ-endotoxin). The potential benefits and risks of herbicide-resistant crops (GRCs) are discussed in this section.

Possible Reduced Use of Herbicides

Proponents have argued that reduction of herbicides adopted for HRC crops occurs primarily because these "new" herbicides are needed in lower doses (if compared for instance with atrazine, 2,4-D, and alachlor) and are applied later in crops, post-emergence. However, higher resistance of the crop to the target herbicide would, in practice, suggest to the farmer to adopt a higher rate than advised to ensure that all weeds are burned in one tractor trip with the targeted broad spectrum herbicide.

Improved Integrated Pest Management (IPM)

Integrated pest management IPM could benefit from some HRCs if alternative non-chemical methods were applied first to control weeds and the target herbicide were used later, only when and where the threshold of weeds is surpassed, in postemergence. If HRCs could be implemented in IPM programs without underestimating the induction of weed resistance and adopting all the available nonchemical alternatives to manage weed control, this technology would be a step toward more sustainable agriculture. However, in practice, insufficient work of extension outreach and appropriate protocols promoted by the producers could only lead to a further link of the farmland to the producers and their marketing policies aimed at increasing their sales of the targeted herbicides aside to their HRCs seeds.

TABLE 1.2 : HERBICIDE-RESISTANT CROPS (HRCS) APPROVED FOR FIELD TESTS

Crop	Herbicide	Research organization
Alfalfa	Glyphosate	Northrup King
Barley	Glufosinate/Bialaphos	USDA
Canola (oilseed rape)	Glufosinate/Bialaphos	University of Idaho
		Hoechst–Roussel/AgrEvo
	Glyphosate	InterMountain Canola
		Monasnto
Corn	Glufosinate/Bialaphos	Hoechst–Roussel/AgrEvo
		ICI
		Upjohn
		Cargill
		DeKalb
		Holdens
		Pioneer Hi-Bred
		Asgrow
		Great Llkes Hybrids
		Ciba–Geigy
		Genetic Enterprises
	Glyphosate	Monsanto
		EdKalb
	Sulfonylurea	Pioneer Hi-Bred

Table 1.2 Contd.

Crop	Herbicide	Research organization
		Du Pont
	Imidazolinone	American Cyanamid
Cotton	Glyphosate	Monsanto
		Daiyland Seeds
		Northrup King
	Bromoxynil	Calene
	Monsanto	
	Rhone Poulence	
	Sulfonylurea	Du Pont
		Delta and Pine Land
	Imidazolinone	Phytogen
Peanuts	Glufosinate/Bealaphos	University of Florida
Potatoes	Bromoxynil	University of Idaho
		USDA
	2,4-D	USDA
	Glyphosate	Monsanto
	Imidazolinone	American Cyanamid
Rice	Glufosinate/Bialaphos	Louisiana State University
Soybeans	Glyphosate	Monsanto
		Upjohn
		Pioneer Hi-Bred
		Northrup King
		Agri-Pro
	Glufosinate/Bialaphos	Upjohn
		Hoechst/AgrEvo
	Sulfonylurea	Du Pont
Sugar beets	Glufosinate/Bialaphos	Hoechst–Roussèl
	Glyphosate	American Crystal Sugar
Tobacco	Sulfonylurea	American Cyanmid
Tomatoes	Glyphosate	Monsanto
	Glufosinate/Bialaphos	Canners Seed
Wheat	Glufosinate/Bialaphos	AgrEvo

Benefits to Developing Countries

Although the majority of HRCs currently on the market and under development belong to key crops in Western agriculture, a few innovations have been proposed that would help developing countries. For example, HRCs have been proposed for improved control of parasitic flowering seeds such as *Orobanche* and *Stringa*, both which can severely reduce grain yields. The HRCs would permit more effective herbicide action against the soil parasitic weed without damaging the target crop. Trials on boomrape have demonstrated that the engineered plants, can overproduce at a rate at least double the reproduction of the control plants. However, the authors observed that this technology can only be used with weeds that do not have the potential to interbreed with wild relatives that could themselves become weeds. For example, in northern African countries, most crops such as sorghum, wheat, and canola (oilseed rape) have their wild relatives nearby, which therefore increases the risk that genes from the herbicide-resistant crop varieties can be transferred to wild relatives. The same gene escape risks are possible for tomato, corn, and potato in South America.

POTENTIAL RISKS

The risk that herbicide-resistant genes from a transgenic crop variety can be transferred via pollination into weedy relatives has been demonstrated for canola (oilseed rape) and sugar beet. Mikkelsen et al. (1996) and Brown and Brown (1996) have shown that herbicide-resistant genes from a transgenic canola move quickly into wild relative weedy populations. Boudry et al. (1994) also noted consistent gene flow between the cultivated sugar beets and weed beet populations.

Repeated use of the same herbicide in the same area creates problems of plant resistance to the target herbicide. This concern has consistent bases in the recent history of herbicides. For instance, if glyphosate from an actual few million hectares of crops were allowed to associate with HRC crops, the resulting acreage could be around 70 million ha, and pressure on weeds to evolve resulting acreage could be around 70 million ha, and pressure on weeds to evolve resistant biotypes could be pronounced. Sulfonylureas and imidazolinones, to be targeted in HRC crops, are particularly prone to rapid evolution of resistant weeds and have already resulted in several cases of resistance. Extensive adoption of HRCs will increase the acreage and surface treated, thereby exacerbating the resistance problems.

Environmental Risks

Even if less environmentally persistent than previous herbicides (e.g., Alachor, 2,4-D, atrazine), the "environmentally friendly" HRCs have, as do most chemical pesticides, consistent or severe environmental effects.

Bromoxynil (Commercial Name: Buctril)

Bronmoxynil has been targeted in HRC cotton by Calgene and Monsanto. This herbicide has traditionally been used in winter cereals, cotton, corn, sugarbeets, and onions to control large-leaf weeds. Drift has been observed that has resulted in damage to nearby grapes, cherries, alfalfa, and roses. In addition, leguminous plants can be very sensitive to this herbicide, and potatoes can be damaged as well. Consistent residues over the accepted standards have been detected in soil and groundwater and as fallout. Rodents tested have demonstrated some mutagenic responses. Stafilinid beetles have been shown to have reduced survival and egg production at suggested dosages. Crustaceans *(Daphnia magna)* have also been severely affected.

Glufosinate/Bialaphos(Commercial Name: Basta)

Many crops have been modified for this herbicide-resistance PAT (phosphinothricin acetyl transferase) gene, which has been introduced into alfalfa, corn, barley, wheat, rice, canola, peanuts, soybeans, sorghum, tomatoes, and sugarbeets (Table 1.2). Turfgrass *(Agrostisis stolonifera)* and other components have been engineered for resistance but need appropriate environmental risk assessement before being marketed. This herbicide has been on the market since 1984 as a synthetic development of a natural pathogen toxin of *Streptomyces viridochromogenes*. The amount of active ingredient used is 200–900 g/ha postemergence on the engineered plants.

Although detrimental effects on users and consumers seem unlikely under recommended doses, toxic effects on humans and animals have been reported. For example, the Basta surfactant (sodium polyoxyethylene alkylether sulfate) has been shown to have strong vasodilative effects in humans and cardiostimulative effects in rats. Treated mice embryos exhibited specific morphological defects.

Imidiazolinone (Commercial Name: Imazethapyr)

Imidiazolinone is applied at low doses (38–50 g/ha) in beans and soybeans postmergence. It has been observed that, at the field rate of 50 g/ha, there is no effect in laboratory and field microbial biomass. However, higher doses induce catalase and hydrolase activity and increase the risk for monocultural practices and reduced mycelial growth in *Sclerotinia trifoliorum.*

Sulfonylurea (Commercial Name: Safari)

Sulfonylurea is used as a herbicide on wheat, barley, sugarbeets, cotton, maize, potatoes, and soybeans postmergence. Drift from very low amounts (5–30 g/ha) can damage cultures, and potential losses may result in several crops, wild plants, and nontarget invertebrates.

Glyphosate (Commercial Name: Roundup)

Most HRCs have been engineered for glyphosate resistance. Although adverse effects of herbicide-resistant soybeans have not been observed on feeding animals such as cows, chickens, and catfish; genotoxic effects have been demonstrated in other nontarget organisms. Earthworms have been shown to be severely damaged by the glyphosate herbicide at 2.5–10 1/ha. For example, *Allolobophora caliginosa,* the most common earthworm in European, North American, and New Zealand fields is damaged by this herbicide. In addition, aquatic organisms, including fish, can sometimes be severely damaged. The prevalence of the nematode *Steinerema feltiae,* a useful biological control organism, is reduced by 19–30% by use of this herbicide.

Health Risks

The unknown health risks associated with the use of herbicides (as well as most xenobiotics) involve the effects of low-level chronic exposures. Although most research has addressed cancer risk, much less research has been focused on neurological, immunological, developmental, and reproductive effects. Much of this problem reflects the dearth of methodologies and diagnostic testes at the disposal of scientists necessary to evaluate the risks caused by exposure to many chemicals, including herbicides, properly.

Although industry often stresses the desirable characteristics of their HRCs, environmental and alternative agriculture groups, as well as some scientists, disagree, with the contention that these products are safe. For example, research has shown that the application of glyphosate can increase the level of plant estrogens in the bean *Vicia faba*. Feeding experiments have shown that cows fed transgenic glyphosate-resistant soybeans had a statisically significant difference in daily milk production compared with other test groups.

INCREASED COSTS ASSOCIATED WITH USE OF HRCs

Because the herbicides for which HRCs are being designed are almost all under patent, they will be more expensive than many of the herbicides they are intended to replace. In addition, although analysts project that, for example, switching to bromoxynil for broadleaf weed control in cotton could result in savings of $37 million each year from reductions in herbicide purchases of 40% to 50%, few economic product evaluations that demonstrate cost savings with the use of HRCs have been published. Furthermore, recent problems with use of glyphosate-resistant cotton in the Misissippi Delta region (crop losses resulting in up to $5000 thousand of this year's cotton crop) suggest this technology needs to be perfected further before some farmers will reap economic benefits.

Some scientists suggest that use of HRCs will cause a shift to the use of fewer broad-spectrum herbicides, which will reduce the amount of sprays and the amount of herbicide used per application, others contend that the overall use of HRCs will actually increase herbicide use and thereby increase costs associated with their application.

BACILLUS THURINGIENSIS (BT) FOR INSECT CONTROL

More than 40 BT crystal protein genes have been sequenced and 14 distinct genes identified and classified into 6 major groups based on amino acids and insecticidal activity. Many crop plants have been engineered with the BT δ-endotoxin, including alfalfa, corn, cotton, potatoes, rice, tomatoes, and tobacco (Table 1.3). The amount of toxic protein expressed inside the modified plant is 0.01–0.02% of the total soluble proteins.

Potential Economic and Environmental Benefits

Supporters of BT usage cite current trials demonstrating a high level of efficacy in controlling corn borer damages on plants. Corn engineered with BT δ-endotoxin has the potential to recover 5–15% of corn borer damage in 70 million acres in the United States with a projected benefit of $50 million annually. However, with out careful management tactics based on university agriculture extension and technical expertise, growers and seed suppliers could see unforseen environmental risks.

TABLE 1.3 : TRANSGENIC INSECT-RESISTANT CROPS CONTAINING BT ENOTOXINS. APPROVED FIELD TESTS

Crop	Research organization
Alfalfa	Mycogen
Apples	Dry Creek
	University of California
Corn	Asgrow
	Cargill
	Ciba–Geigy
	Dow
	Genetic Enterprises
	Holdens
	Hunt–Wesson
	Monsanto
	Mycogen
	NC+Hybrids
	Northrup King
	Pioneer Hi-Bred
	Rogers NK Seed
Cotton	Calgene
	Delta and Pineland

Table 1.3 Contd.

Crop	Research organization
	Jacob Hartz
	Monsanto
	Mycogen
	Northrup King
Cranberry	University of Wisconsin
Eggplant	Rutgers University
Poplar	University of Wisconsin
Potatoes	USDA
	Calgene
	Frito-Lay
	Michingan State University
	Monsanto
	Montana State University
	New Mexico State University
	University of ldaho
Rice	Louisiana State university
Spruce	University of Wisconsin
Tobacco	Auburn University
	Calgene
	Ciba–Geigy
	EPA
	Mycogen
	North Carolina State University
	Roham & Haas
Tomatoes	Campbell
	EPA
	Nonsanto
	Ohio State University
	PetoSeeds
	Rogers NK Seeds
Walnuts	University of California, Davis
	USDA

Cotton

Cotton was the first crop plant engineered with the BT δ-endotoxin released into the market. It has been estimated that caterpillar pests, including the cotton bollworm and budworm, cost U.S. farmers about $171 million/year as measured in yield losses and insecticide costs. Benedict et al. (1992) predicted that the widespread use of BT-cotton could reduce insecticide use and thereby decrease costs by as much as 50 to 90%, which would save farmers $186 million/year, respectively.

Potatoes

Genetically engineered Russet Burbank potatoes with *Bacillus thuringiensis tenebrionis* (the δ-endotoxin is represented inside the engineered potatoes at 0.05–0.1% CryIIIA as a percentage of total protein) have been shown to be successful. This formulation is also suggested as a base to develop an effective and sustainable IPM program for potatoes.

TABLE 1.4 : THE NUMBERS OF NATIVE AND INTRODUCED PLANTS AND INSECTS

Type	Plant	Insects
Native species	2,525	11,512
Immigrant species and established in nature	0	946
Introduced species and established in nature	925	42
Cultivated species but not present in nature	25,000	5

Eggplants

The δ-endotoxin Cry3b-engineered eggplants have demonstrated consistent effects against potato beetles. However, resistance is one harrowing prospect, and proper resistance management has to be considered.

Potential Environmental Risks

Insects that develop resistance to transgenic crop varieties are one of the posssible risks associated with the use of BT δ-endotoxin in genetically engineered crop varieties. Resistance to BT has already been demonstrated in the cotton budworm and bollworm. If BT -engineered plants become resistant, a key insecticide that has been utilized successfully in IPM programs, could be lost. Therefore, proper resistance-management strategies with the use of this new technology are imperative. Another potential risk is that the BT δ-endotoxin could be harmful to nontarget organisms. For example, it is not clear what potential effect BT δ-endotoxin residues that are incorporated into soils will have against an array of nontarget useful invertebrates living in the rural landscape.

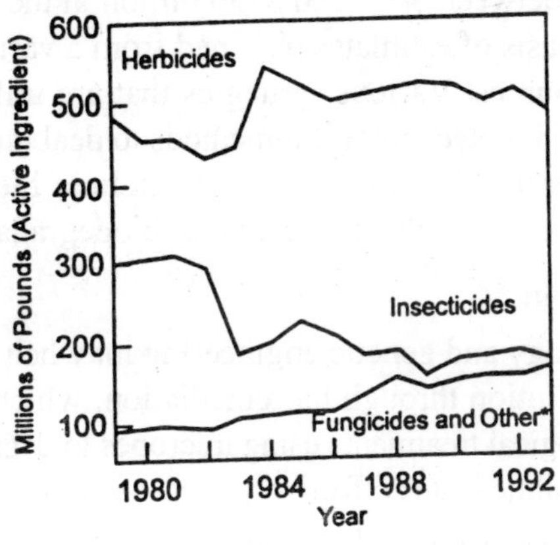

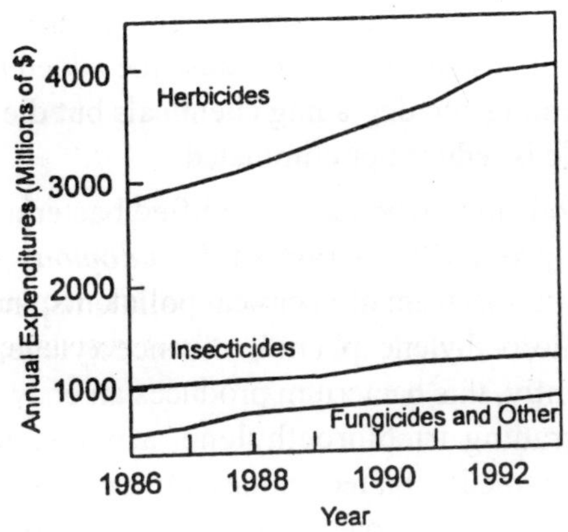

Figure 1.1 : ***The choice of herbicide resistance as the main target for engineered crops seems related to the promising trend of herbicide sales and returns rather than to other environmental strategies such as potential reduction of pesticides in the environment.***

RECYCLING OF TOXIC WASTES

Advances in technology have resulted in the availability of a variety of chemicals, many of which have increased ecosystem pollution. Currently, in the United States, some 70,000 different chemicals are released into the environment through soil, water, and air; an estimated 100,000 chemicals are used worldwide. Cleanup of hazardous wastes by conventional technologies is projected to cost between $400 and $750 billion in the United States alone on the basis of estimates obtained from a variety of federal and private sources. Various strategies that are utilizing genetic engineering and biotechnological methods to deal more efficiently with waste removal and management include bioremediation, phytoremediation, and the production of biodegradable plastics.

Bioremediation

Biotechnology and genetic engineering may help reduce environmental pollution through bioremediation, which is the application of biological treatments using microbes to degrade chemical materials at polluted sites effectively. Because bioremediation provides continuous cleanup of contaminated sites, such as pesticide residues in agricultural ecosystems, it has significant advantages over other techniques. Furthermore, a marked degree of self-regulation is present in such systems because the added microbes survive by consuming and degrading chemicals but die off when the nutrient source is reduced or eliminated.

Investigations into genetically modified bacteria for bioremediation have produced a strain of *Pseudomoas cepacia* that degrades a wide spectrum of chemical pollutants, including vinyl chloride, dichloroethylene, phenol, toluence xylene; and creosol. Most importantly, this bacterium produces an enzyme that is capable of degrading trichloroethylene, a persistent industrial degreaser. Bioremediation techniques are growing in importance and diversity as new approaches are developed to use both wild and genetically engineered biota for chemical pollution abatement.

Phytoremediation

Phytoremediation, or the use of specially selected and engineered metal-accumulating plants for environmental cleanup, is an emerging technology that may serve as a cost effective approach to treating soils and groundwater contaminated with toxic metals. Currently, cleanup of sites in the United States contaminated with heavy metals and organic mixtures can alone cost $42.5 billion. Phytoextraction, or the use of metal accumulating in plants to remove toxic metals from soils, could be utilized to clean up sandy loam soil to a depth of 50 cm, which would cost $60,000–$100.000 compared with at least $400,000 for excavation and storage alone using traditional soil removal methods. Rhizofiltration, or the use of plant roots to remove toxic metals from polluted waters, may offer an advantage in water treatment because of the ability of plants to remove 60% of their dry weight as toxic metals, thus markedly reducing the generation and disposal costs of hazardous radioactive residue. Rhizofiltration could also be a cost-competitive technology in the treatment of groundwater containing low concentrations of toxic metals.

Genes encoding the cadmium-binding protein, methallothionein, have been recently shown to be expressed in plants. A research group at Peking University has engineered a human gene encoding methallothionein (a protein that binds heavy metals) into tobacco. The genetically engineered plants have survived exposure to cadmium concentrations that were 25 times greater than the dose that killed control plants. More importantly, these genetically engineered plants have also been shown to absorb cadmium from the soil. Researchers are also now engineering the gene into weeds in the hope of using the transgenic weeds to reduce heavy metal contamination on farmland.

Biodegradable Plastics

Another area of waste management biotechnology and genetic engineering could play a role is in the production of biodegradable plastics. Currently, plastics account for 20% by volume of all municipal solid wastes. Estimates of the current global market for biodegradable plastics range up to 1.3 billion kg per year. However, until recently, the cost of these biodegradable plastics, such as polyhydroxyalkanoates (PHAs), polymers made entirely by bacterial fermentation, has been one of the major limiting factors in their use. Producing PHAs in plants would significantly reduce the expense of manufacture compared with fermentation and make these biopolymers competitive with petroleum based plastics for low-cost uses.

Recently, efforts to produce PHAs in plants have been successful, Bacterial genes that are necessary to synthesize polyhydroxybutyrate (PHB), a type of PHA, were transferred to *Arabidopsis thaliana* plants. The genetically engineered plants accumulated up to 14% dry weight of PHB, without deleterious effects on plant growth or fertility and with a level of polymer yield that is considered commercially practical. Several companies are now pursuing development of PHA-producing genetically engineered oilseed rape (canola) crops.

Environmental Risks

In initial prelease testing, the interactions of such new genetically engineered organisms with nontarget organisms in the soil communities and contaminated landscapes must be carefully monitored to avoid potentially deleterious side effects. Also, it will be important to monitor the safe environmental disposal of plants that accumulate toxic materials.

NOVEL FOOD PRODUCTS

The objective of biotechnology and genetic engineering is to improve the food supply by increasing crop and livestock productivity, enhancing nutrient composition and availability, and improving food characteristics such as size, taste, and texture. The availability, and improving food characteristics such as size, taste, and texture. The following are examples of the types of applications that have been achieved or are under way.

Increased Crop and Livestock Productivity

In the past, scientific breeders of plants and animals have utilized the rich genetic resources of cultivars and land races in crop improvement programs. It has been estimated that at least half of the increase in agricultural productivity realized this century may be directly attributable to "artificial selection, recombination, and intraspecific gene transfer procedures". Use of biotechnology and genetic engineering may result in even greater productivity gains through high-yield crop varieties, improved pest and disease control, the production of herbicide-resistance in crops, enhanced nutrient availability, and tolerance to a variety of environmental stressors.

Biotechnology is also being used as a way to increase livestock productivity by enhancing milk and meat production. This is achieved by isolating the gene for the protein hormone (e.g., somatotropin) made in the anterior pituitary gland of animals and inserting it into bacterial cells, which results in an increased production of the hormone. Applications of this technology include the injection of recombinant bovine somatotropin (rPST) into dairy cows to increase milk production and injection of recombinant porcine somatotropin (rPST) into growing pigs to increase carcass leanness and decrease fattiness. Some authors have speculated that decreasing the quantities of feed consumed per unit of output may benefit the environment by decreasing environmental pollution through reducing the quantity of feed provided; reducing the quantity of fertilizer and other inputs associated with growing, processing, and storing animal feed; and reducing animal waste products.

Biotechnoligy may also play a role in improving crop plants that, in turn, enhance secondary productivity in livestock. Researchers at the Common wealth Scientific and Industrial Research Organization's division of plant industry in Canberra have found a way to insert a sunflower gene into the subterranean clover, a major constituent of Australian sheep pastures. The new clover, with genes from sunflower that code for a protein in sulfure amino acids, provides sheep with a sulfurrich diet that is necessary for wool production. Creating such a transgenic clover resulted in about a hundredfold increase in the sulfure-rich protein. However, it has been estimated that (i.e., an increase in wool production by 5–10%), the level of sulfur-rich protein needs to be boosted another tenfold to have a substantial impact on wool production. The group is testing gene promoters that could achieve this. Producing the same amount of wool from fewer sheep could reduce soil erosion and other environmental pollution produced by these animals.

Developments in biotechnology and genetic engineering may also be able to improve ruminant nutrition by modifying the microbes that are involved in ruminal fermentation. The objective will be to find suitable foreign bacterial genes that can be inserted into ruminal bacterial organisms. Techniques of genetic engineering are also playing an important role in increasing animal productivity by improving and developing vaccines and pharmaceuticals (e.g., fertility hormones). Hybridoma technology, which results in the generation of monoclonal antibodies by cell fusion procedures, will be increasingly useful in diagnosing specific diseases as well as in disease prevention and treatment. The broad range of potential vaccines for control of various diseases is especially promising because of their low environmental risks and excellent socioeconomic benefits.

Enhanced Nutrient Composition

Another application of biotechnology in agriculture is intended to improve the food supply by enhancing the nutritional composition of foods. In 1992, Monsanto was able to produce a genetically engineered potato successfully with an increased starch content. A higher starch content reduces oil absorbtion during frying and thereby lowers the cost of frying certain food products (e.g., French fries, potato chips) and reduces oil content in the finished product. Genetically engineered strains of oilseed rape (canola) are undergoing development at Calgene, where applications include high-stearic-oil-content margarine and edible canola oil with reduced saturated fat content.

Through biotechnology, scientists also hope to create foods that protect against cancer, heart disease, osteoporosis, and other life-threatening illnesses. For example, the National Cancer Institute (NCI) is currently working on increasing the amount of phytochemicals, which are linked to cancer prevention, in foods such as garlic, parsley, and citrus fruits. Government, university, and industry scientists are also developing cereal grains with increased amounts of both soluble and insoluble fiber to lower cholester[illegible]d fight digestive cancers, respectively; milk with improved calcium bioavailability to help protect against osteoporosis; and vegetables with boosted levels of antioxidants (e.g., super carrots with five times the amount of β-carotene) to help reduce the risk of cancer.

Enhanced Nutrient Availability

Biotechnology and genetic engineering can also be employed to enhance nutrient availability in agricultural systems. Genetic engineering has been applied to the problem of nitrogen fixation with a specific focus on the genetic makeup of organisms that fix nitrogen from the atmosphere and the genetic basis for the relationship between leguminous species (e.g., peas, beans, alfalfa, clover, peanuts) and the nitrogen-fixing bacteria that occupy their nodules. Scientists in China have recently developed recombinant strains of nitrogen-fixing bacteria that have a higher nitrogen fixation efficiency than traditional bacteria. These recombinant strains of bacteria have been spread over more than a million hectares of rice and soybean fields, and preliminary results show crop yield increases of 5–10%.

In addition, a mechanism of biological nitrogen-fixation similar to that of natural legumes eventually may be genetically engineered into plants such as wheat and corn. If this is achieved, the need for commercial nitrate fertilizers could be significantly reduced. However, because the molecular mechanisms required for symbiotic nitrogen fixation are complex (involving at least 17 genes), achieving this will require an investment in research over several decades.

Other Improved Food Characteristics

Other applications of biotechnology to improve the food supply include the improvement of certain food characteristics such as size, ripeness, acidity or sweetness, taste, and texture. Genetic modifications have enabled the production of fruit that may have better taste and an enhanced shelf life through delayed pectin degradation or altered responses to the plant hormone ethylene. Among the first such novel products on the market was Calgene's slow-ripening Flavr Savr® tomato, which was approved for sale by the Food and Drug Administration in May 1994. This tomato has been transfected with an "antisense" gene responsible for the enzyme polygalacturonase, which solubilizes pectin. Because pectin degradation increases fruit ripening and decreases shelf-life, preventing translation of the message (by gaving the antisense message present) for polygalacturonase should retard fruit ripening and increase prepurchase and postpurchase shelf life (AMA 1991). DNA Plant Technology (DNAP) corporation's Endless Summer® tomato, which like Calgene's Flavr Savr has an antibiotic-resistance marker gene encoded into it, also incorporates a gene-splicing technique to retard ripening.

RISKS

Environmental Risks

The production of plants and animals to suit specific environments could lead to the transformation of more land presently occupied by natural ecosystems into agricultural land. The ecosystems of the tropics on the land not suited for agriculture of any kind are particularly vulnerable. Thus, the removal of yet more natural ecosystems would further deplete biodiversity. Genetic engineering may also favour monocultures that will threaten the global centers of ordered biodiversity.

Health risks

One potential health risk is related to the use of genetically engineered animal hormones. One can illustrate using the case of rBST. The Food and Drug Administration has ruled that the presence of rBST in milk is safe for children and adults (FDA 1995). Even so, there have been questions regarding the impact of this technology on animal and human health. Use of rBST in dairy cows increases the chances of bacterial infections and mastitis and also reduces the reproductive cycle in treated dairy cows. Milestone et al. (1994) report that increased infections in cattle will require treatment with antibiotics. Although not all antibiotics appear in milk, some do. Thus, if more antibiotics are used, an indirect risk to humans may arise because some residues may remain in the milk.

A second potential human health concern is the risk of introducing allergens into the food supply. A recent study by Nordlee and colleagues demonstrated the transfer of a major food allergen from Brazil nuts to transgenic soybeans during the development of a genetically engineered crop variety. Furthermore, although only a dozen foods may produce allergic reactions— mainly protein foods— biotechnology allows for nontraditional food proteins (e.g., moths, insects) to be present for which no knowledge currently exists regarding their allergenic or nonallergenic qualities.

A third potential human risk relates to the increasing prevalence of antibiotic-resistant and disease-causing bacteria. Because antibiotic-resistant genes are the most commonly used type of selectable marker in genetic engineering and are rarely deleted from the resulting organism, incorporation of antibiotic-resistant markers into the genetic material of human pathogens could pose risks by increasing the prevalence of antibiotic-resistant and disease-causing bacteria. In 1989, dozens of people died and thousands others were crippled after consuming a batch of synthetic L-tryptophan will never be known, it has been reported that a specific impurity stemming from the bacterial strain may be cause of the syndrome.

A fourth health concern is that genetically engineered crops may be able to transfer their foreign gene(s) to other unrelated micro-organsims. Genetically engineered oilseed rape (canola), black mustard, thorn-apple, and sweet peas all contain an antibiotic-resistance gene and were grown together with fungus *Aspergillus niger*, or their leaves were added to the soil. The fungus was shown to have incorporated the antibiotic-rsistance gene in all coculture experiments.

Other health risks created through the use of biotechnology and genetic engineering include the ability of genetically engineered organisms to survive and harm nontarget organismsm the creation of new toxic organisms, or both. The risk of a new host being infected by a virus or recombining to form a more deadly, virulent virus must be investigated further. For example, the risk of recombination between the engineered vaccine virus and other orthopox viruses endemic in wildlife, such as the cowpox virus, still needs to be investigated accurately.

Social and Economic Impact

Proponents often cite biotechnology and genetic engineering as the way to help solve the world's food problems (i.e., improve agriculture in developing countries). However, some critics who are skeptical about of ability of biotechnology and genetic engineering to increase food production project there will only be about a 1 % increase in crop yield on the basis of the anticipated contributions of biotechnology during the next two decades. In addition, no increase in crop yields is projected for Africa because no major advances in biotechnology are expected to be applied in Africa in the near future.

Some persons further question why, after 20 years of research, biotechnologists have not produced a single high-yielding variety of wheat, rice, or corn. The answer, according to plant scientists, is that plant breeders using traditional techniques may have largely exploited the genetic potential for increasing the share of photosynthate that goes into the seed. Others feel that the present channeling of funds to expensive biotechnology projects diverts scarce resources from other research that could focus on more practical solutions to pressing social problems such as hunger and food insecurity.

Some authors have also observed that many of the crops being engineered for herbicide resistance belong to the group of key crops in Western agriculture. This circumstance may reflect the domination of the majority of the biotechnology industry by transnational companies (TNC) in the developed world whose business it is to generate profits. If sustaining the Third world is the target of genetically engineered crops, other vegetables and crops have to be considered. Also, helping these countries bypass expensive, high-input crop production and move their traditional agriculture toward low-input sustainable practices is desirable as well.

TRANSGENIC ANIMALS

Benefits

Among the benefits of producing genetically engineered or

transgenic animals are the ability to manufacture more cost-effective drugs (including vaccines), new animal models to study human disease, human tissue and organ harvesting, and improvements in the food supply (e.g., modifying the shape, size, or nutritional quality of animals). In the early 1980s, the first successful experiments creating transgenic vertebrates were reported. Palmiter et al. (1982) created a transgenic mouse by transferring a growth hormone into the embryo of the mouse. In 1988, a patent was awarded for a mouse that was genetically engineered with human genes designed to be tumorigenic. This mouse, commonly referred to in the scientific literature as the "oncomouse" (oncology is the study of tumors), prompted the first patent issued for a transgenic animal (OTA 1989).

Transgenesis, or the transfer of genes across species lines, opened the way for animals to be used as an alternative to tissue-culture productions of human protein. In 1987, a transgenic mouse was created that demonstrated the viability of tissue-specific expression of foreign proteins. As reported by Krimsky and Wrubel (1996), "The mouse was genetically engineered to make the clot-dissolving factor tissue plasminogen activator (TPA), which is viewed as a highly promising drug for the treatment of coronary heart disease and a strong competitor of the widely acclaimed streptokinase."

Krimsky and Wrubel (1996) also note, "Finnish researchers have developed a genetically-modified cow that purportedly can produce mild containing large amounts of erythropoietin (red cell growth factor) used to treat anemia. It successful, this method will replace the costlier cell culture techniques. Other more remote applications of transgenic animals include human blood and organ production."

Trangenic animals under development include swine with the human growth hormone gene, genetically engineered livestock desinged to tolerate extreme climatic conditions, transgenic sheep that grow faster than normal sheep, engineered sheep that secrete insect repellent and produce moth-proof wool, and genetically engineered sheep and cows that produce milk consumable by lactose-intolerant individuals. Many different species of fish are being genetically engineered to increase fish size and growth rate or improve survival in new environments. Genetic engineers have turned much of the attention toward finfish and shellfish.

Many species of transgenic fish have been grown in the laboratory. Fast-growing Pacific salmon have been engineered by various groups of researchers worldwide. Scientists attached a switch to a growth hormone gene from coho salmon and injected the transgene into chinook salmon eggs. On average, the transgenic salmon grew to be elevenfold heavier than their age-mates. Fast-growing fish have also been produced outside the laboratory. In 1991, transgenic carp were tested in a high security pond at Auburn University. These fish were fitted with a growth hormone gene from rainbow trout that enabled them to grow 40% faster than normal. Researchers have also identified a protein in winter flounder, which has recently been transfered into Atlantic salmon, that prevent fish blood from freezing. Transgenic salmon with this trait could potentially be raised in sea pens farther north, where the species could not otherwise live.

RISKS

Environmental

Very real environmental risks may be associated with transgenic animals—particularly with certain applications, such as transgenic fish. For example, transgenic fish have the potential to disturb ecosystems seriously by gaining a competitive advantage in the wild ecosystem. A fast-growing transgenic fish could assume a higher than usual position if the food chain because of its greater size and ability to compete for food, which could harm native species, or a freeze-tolerant transgenic fish raised in the north could escape into a geographic area from which it was previously excluded, and cause competition that would harm native species. Fertile transgenic fish could also successfully invade ecosystems, thus exacerabating the present problem with exotic invaders in aquatic ecosystems.

Health

Using transgenic animals to harvest blood, tissues, or organs may create certain health risks. For example, take the case of deriving human hemoglobin from transgenic pigs. Human hemoglobin must be separated from that of the animal to ensure the purity of the product. Also, how humans will respond to such animal-derived human hemoglobin is still untested.

Social and Ethical

Finally various social and ethical risks are associated with the production and use of transgenic animals. Certain animal rights groups (e.g., The Humane Society) question whether it is morally or ethically correct to turn animals such as mice, pigs, and sheep—which unlike plants and bacteria are sentient beings—into biomachines for the manufacture of proteins and other biological materials (Fox 1992). Some groups may also find it unethical to introduce human genes into livestock and plants. This application of genetic engineering could raise ethical concerns and seriously undermine the public's perceptions of biotechnology. Religous groups may also have certain conflicts as to whether this new technology is compatible with each of their traditional norms. There may also be a conflict of interest within the value and belief systems of those groups for whom it is abhorrent to manipulate, control, experiment with, or consume animals of any type.

GENERAL RISKS OF RELEASING GENETICALLY ENGINEERED ORGANISMS INTO THE ENVIRONMENT

Single-Gene Changes and Pathogenicity

Most single-gene changes are probably not likely to affect the pathogenicity and virulence of an organism in nature adversely (NAS 1987). However, some gene changes may have detrimental consequences. Certain genetic alterations in animal and plant pathogens, for example, have led to enhanced virulence and increased resistance to pesticides and antibiotics. For instance, some oat rust microbes, initially nonpest genotypes for a particular oat variety, became serious pest genotypes after a single gene change allowed the rust to overcome resistance in the oat genotypes after a single gene change allowed the rust to overcome resistance in the oat genotype.

An important fungal disease of rice, rice blast, has been demonstrated to have genotypes with single-gene changes that cause the fugnal organism to be potentially pathogenic to rice cultivars. A similar phenomenon of single gene changes resulting in pathogenicity has been documented with a related fungal pathogen that infects weeping love grass. This phenomenon has led plant pathologists to develop the "gene-for-gene" principle of parasite–host relationships in which a single mutation in a parasite overcomes single-gene resistance in the host. Furthermore, numerous instances have been documented in which insects, through a single, gene change, have overcome resistance in plant hosts or have evolved resistance to insecticides. More than 500 species of arthropods have developed resistance to pesticides.

Threats from Modified Native Species

Lindow (1983) has reported that there is little or no danger from the ice-minus strain of *Pseudomonas syringae* (Ps) because Ps is a native U.S. species that produces related phenotypes in nature. Other investigators have demonstrated that there are different genotypes of Ps, and some of these genotypes have genes for pathogenicity. Because some native species have the ability to alter their interactions within an ecosystem, the genetic modification and release of native species into the natural ecosystem may not always be safe. For example, from 60 to 80% of the major insect pests of U.S. and European crops, respectively, were once harmless native species in the United States and Europe. Many of the insects moved from benign feeding on natural vegetation to destructive feeding on introduced crops. For instance, the Colorado potato beetle moved from feeding on wild sandbar to feeding on the potato that was introduced from Peru and Bolivia. This insect has become a serious pest of the potato in the United States and Europe.

Intentional Introduction of Crop Plants and Animals

Some proponents of biotechnology suggest that the intentional introduction of foreign plants and animals into the United States is a good model for predicting potential problems arising from biotechnology (NAS 1987). If so, there is reason for concern because several serious problems have resulted from the intentional introduction of what were believed to have been beneficial crops and animals. Genetic similarities between many of the crops and weeds are evident from the fact that 11 of the 18 most serious weeds of the world are crops in other regions of the world. Of the several thousand crops that were intentionally introduced into the United States, 128 species of agricultural and ornamental plants have serious pest weeds. Some of these introduced plant species. like Johnson grass, are among the most serious weed species in the United States—especially in the southeastern United States. Johnson grass was introduced as a forage for livestock before it escaped and became a weed pest.

This pattern of native species and introduced plants species in the United States is not unique. For example, Florida has only 2525 indigenous plant species, but approximately 25,000 plant species have been introduced and are under cultivation there.One exotic pest that is displacing native plant species is the melaleuca (*Melaleuca quinquenervia*) tree introduced from Australia. This soclearly illustrates the threat to our natural ecosystem when so-called beneficial organisms are introduced and released into nature. The picture in Florida for insects is quite different than for plants.

The great majority of insects species (11,512) are native, with nearly 1000 immigrant species. A relatively small number (42) of insect species are established in nature. One of these species that has established itself and is a serious pest in nature is the imported fire ant. Furthermore, 9 out of 20 introduced domestic animal species in the United States have displaced or destroyed native species. These introduced domestic animals, including donkeys, horse, and goats, have become serious ecological pests. A total of 10 other introduced animals, including mammals (e.g., mongoose and wild boars) and birds (e.g, English sparrow and mynah), have become pests.

Furthermore, at least 70 species of fish have been introduced and have become established in U.S. aquatic ecosystems. These 70 species represent about 10% of all U.S. fish species. A total of 5 introduced fish species have become pests, displacing and reducing the number of native species and, in other cases, altering the habitat and making it uninhabitable for fish and other species. In addition to these intentionally introduced fish, there is concern about the introduction if transgenic fish and the potential ecological effects of these engineered fish in aquatic ecosystems in the United States. This overall history suggests that the introduction of many types of foreign organisms in the ecosystem may have major negative impacts on many of the 500,000 beneficial plants and animals in the United States.

Are Ecological Niches Filled?

An estimated 1500 exotic insect species have been introduced and are established in the United States, and a few of these (17%) have become pests or have a negative impact on native species. This observation indicates that few of the niches in natural ecosystems are filled. There is simple opportunity, therefore, for many species to become established in the United States. Thus, although the argument that engineered organisms will not become established owing to competition with native species may apply in a few cases, most often this is not a valid argument.

USES OF BIOTECHNOLOGY AND GENETIC ENGINEERING

Acceptable and potentially sustainable options for agriculture that could be derived from biotechnology and genetic engineering have been put forward by several authors. Some of the desirable areas of development for such technologies that have the potential to benefit agricultural sustainability, the integrity of the natural environment, and the health and safety of society are discussed in the following paragraphs.

Enhancing Crop Resistance to Pests

Approximately 500,000 kg of pesticides are applied each year in U.S. agriculture, and many nontarget species beneficial to the environment are negatively affected. Genetic engineering targeted for pest control could diminish the need for pesticides.

Resistance factors and toxins that exist in nature can be used for insect pest and plant pathogen control. For example, more than 2000 plant species are known to posses some insecticidal activity, and approximately 700 natural substances in bacteria, fungi, and actinomycetes have fungicidal activity. Traits for resistance to different insect pests and diseases already exist in many cultured crops, including corn, wheat, barley, soybeans, beans, apples, grapes, pears tobacco, tomatoes, and potatoes.

Although some resistance characteristics have been reduced or eliminated in commercial crops, they still can be found in related wild varieties, which provide an enormous gene pool for the development of host– plant resistance. For example a wild relative of tobacco that produces a single acetylated derivative of nicotine is reported to be 1000 times more toxic to the tobacco hornworm than cultivated tobacco.

Transferring this toxic gene to nonfood crops, such as ornamental shrubs and trees, may protect them from certain insect pests. In addition, thionins, proteases, lectins, and chitin-binding proteins that are often present in plants, especially in the seeds, help control some pathogens and pest insects in wild plants. For example, it has been shown that a cowpea protease inhibitor found in the cowpea *Vigna* spp. can now be engineered into tobacco. Laboratory trials also indicate that the cowpea protease inhibitor provides protection against the cotton budworm (*Heliothis virescens*), which is a major pest of tobacco, cotton, and maize.

Development of Perennial Crops

At present, the major cereal crops of the world are annuals. The conversion of annual grains to perennial grains by genetic engineering will reduce tillage and erosion and conserve water and nutrients. Such crops will decrease labour costs, improve labour allocation, and, overall, improve the sustainability of agriculture. Energy efficiency in cultivation of perennial cereal crops will be greatly superior to that of annual crops.

Improved Botanical Pesticides

Only limited quantities of botanical pesticides, such as pyrethrums, are now used in developed countries in place of some synthetic pesticides. However, in some, developing countries, including China and India, botanical pesticides such as neem are effectively used. Increasing the effectiveness of neem and other available botanical pesticides by genetic engineering would be an asset to farmers because these substances are relatively effective and safe.

Bioindication Needs for Sustaninable Use of Genetically Engineered Plants

Bioindication is a strategy that adopts and assesses biological units, species, assemblages of species, and ecosystem models to determine the impact of a selected contaminant such as pesticide residues or fertilizers on the environment. This strategy is aimed at using biological nontarget organisms, both in microcosm-modeled and in field areas, to assess the environmental problems created by adopting certain new management techniques in agroecosystems, including genetically engineered organisms. As observed by several authors, little work has been dedicated to assessing the true environmental impacts of genetically engineered crop plants. For example, there are no data on pollinators of engineered plants modified with the BT δ-endotoxin. It would be rather useful to work with the nontarget species linked to the rural landscapes in which the genetically modified crops are expected to be introduced.

PUBLIC PERCEPTIONS OF BIOTECHNOLOGY

Current public opinion polls show that consumers find some applications of biotechnology and genetic engineering more acceptable than others. For example, in a national random telephone survey conducted in the United States by Hoban and Kendall (1992), 66% of respondents considered plant-to-plant gene transfers acceptable, 25% found animal-to-plant gene exchanges acceptable, 40% found animal gene transfers.

A suvery of public attitudes toward biotechnology in the United Kingdom revealed that a large percentage of respondents would accept genetically manipulated foodstuffs under the condition they were confident about testing and that the food product(s) look and taste the same or better. One-half of respondents felt it would be a good thing to use genetic manipulations to solve the food problems in the Third World, whereas relatively few people felt it would be a good thing to use genetic manipulation to provide or improve the food supply in the Western world. A majority of respondents (70%) also felt that it would be a bad thing to use genetic manipulation for products they did not feel were needed. Consumer surveys in the United States also suggest that the public is skeptical and cautious with regard to certain applications of biotechnology in agriculture and food production. Consumer apprehensions center around biotechnology's perceived unpredictability, risks to the environment, and moral and social questions.

In Europe, public attitudes toward genetically engineered foods appear less favourable than in the United States. Recently, the European Commision has approved mandatory labeling of all genetically modified organisms (GMOs). The European community states that the intent of the label is to serve as a source of information for consumers, not as a warning (Institute for Agriculture and Trade Policy 1997). Result from consumer surveys in the United States. Canada, and the United Kingdom also indicate that most consumers are in favour of labeling genetically engineered foods.

USE OF BIOTECHNOLOGY AS A WAY TO PRESERVE BIODIVERSITY

Benefits

Proponents of biotechnology claim that use of biotechnology and its subdiscipline of genetic engineering is needed to renew the momentum of plant and animal genetics and is one of the factors that has contributed to food production gains achieved thus far (Avery 1993). Advocates also claim that biotechnology has the potential to enhance species preservation, increase the value of biodiversity; and promote ecosystem conservation through decreased environmental degradation.

Enhance Species Preservation

The International Board for Plant Genetics (IBPGR) as well as many others has taken up the job of collecting critically valuable germplasm. The IBPGR gene banks now have more than a half a million plant "accessions" representing thousands of varieties and hundreds of different species. Biotechnology has encouraged the collection of genes through germ plasm secondary to its being viewed as a valuable and prosperous activity. Utilizing biotechnology for collecting germ plasm may help ensure the world's biodiversity is maintained by fostering preservation of species that are in danger of becoming displaced by new varieties (IBPGR 1992).

Increase the Value of Biodiversity

Biotechnology may increase the value that is placed on biodiversity through increased returns on investments in research and development of biotechnologies that may generate animal and crop breeds of potential value. The present economic benefits of biotechnology products are significant and are conservatively estimated to be between $2–3 billion/yr in the United States alone. Worldwide, current benefits are about $6.2 billion/yr. Nearly half of the currente conomic benefits relate to agriculture with significant benefits to the pharmaceutical industry. For example, biotechnology, in the form of bioassays, has reduced the time and cost of screening for pharmaceutical and other uses and has thus increase the value of the underlying genetic resources. For this reason, pharmaceutical companies are becoming more interested in the potential biochemical properties of tropical species and varieties for developing new drugs. On the basis of data from Costa Rica, Aylward (1993) estimates that the net private returns to pharmaceutical companies prospecting for biological resources are around $4.8 million per new drug per year. Over 50% of these royalty returns could realistically be allocated to biodiversity protection in Costa Rica.

Conserve Ecosystems through Decreased Environmental Degradation

The improvement of crop and livestock productivity could, in principle, create an environmental advantage. Less land, particularly less marginal land, would need to be cultivated, thereby reducing problems such as soil erosion and desertification and promoting ecosystem conservation. However, this would require more substantial productivity gains than have been achieved thus far with crop cultivars that have been produced through biotechnology may produce high-yield forestry, which could also help preseve rapidly depleting forests. Although not yet achieved. biotechnology could also help promote environmental management by reducing the use of artificial fertilizers and chemicals and fossil-fuel consumption, which leads to environmental and biodiversity destruction.

Risks

On the other hand, biotechnology and genetic engineering may produce various environmental and social and ethical risks that could ultimately lead to further habitat destruction and depletion of genetic resources. These potential risks include further depletion of biodiversity, harm to nontarget organisms, and exploitation of farmers in developing countries by transnational companies.

Environmental

Further depletion of biodiversity

Use of biotechnology could actually lead to the transformation of more land and thus cause the removal of yet more natural ecosystems into agricultural land, which would further deplete biodiversity. For example, the ecosystems of the tropics, where land may not be suitable for agriculture of any kind, may be particularly vulnerable. Encouraging use of certain genetic resources in the production of pharmaceuticals may also lead to a further depletion of biodiversity. Biotechnology and genetic engineering may also favour monocultures and threaten the global centers of ordered biodiversity—the basis for ecological stability, which has already been seriously undermined primarily as a result of global industrialization, urbanization, and overexploitative agricultural practices.

Harm to nontarget organisms

Other environmental risks created through the use of biotechnology and genetic engineering include the ability of genetically engineered organisms to survive and harm nontarget organisms and the creation of new toxic organisms. The genetic engineering of viruses and bacteria could lead to the accidental production of toxic or environmentally harmful strains. Not all ecological experiments are successful with their potential predictions, and such errors of judgment have had negative effects on the environment in the past.

Socioeconomic

Transnational companies (TNCs) may exploit farmers in developing countries

One social concern of biotechnology is that TNCs are building monopolies over transgenic seed production and may eventually disadvantage poor farmers in developing countries. For example, the packaging of seeds with engineered herbicide resistance and the herbicides by agrochemical companies could be viewed in this light. The situation is made even more complex because the majority of the genetic resources, and thus biodiversity, on which biotechnology depends are found in developing countries. A second concern is that because genes extracted from ecosystems in developing nations can be engineered and turned into valuable assets it will be possible for the genes to be patented by developed nations, which will result in developing-world farmers paying for products that origninated from their nation's own resources.

Risk Assessment and Policy Recommendations

If biotechnology is to contribute to sustainable agricultural development, polices must be adopted to ensure that the profit generated by biotechnological research and development is invested in the conservation of the habitats that produced it and that prospecting ventures contribute economic benefits that build technological capacity in the country of origin. An example that illustrates how this can be done is the partnership between Costa Rica's National Biodiversity Institute (INBio) and Merck and Company, Ltd., by which Merck provides INBio with a $1.1 million dollar budget and INBio provides Merck with plant and animal extracts from biologically diverse, undeveloped areas. In addition, INBio also receives a share of the royalties on any products that are ultimately developed. Mexico, Indonesia, and Kenya are establishing similar bilateral agreements.

A second way that biotechnology can contribute to sustainable agricultural development is through adoption of an International Biosafety Protocol commissioned by article 19 of the Convention of Biological Diversity (CBD), which was introduced and opened for signature at the Earth Summit meeting in Rio de Janerio in 1992. Because genetically modified organisms (GMOs) are new and may pose novel risks, their future impact on the environment and health is uncertain. Many scientists and organizations, including the authors of this chapter, believe the international community would benefit from a legally binding protocol that sets basic standards for the release and export of GMOs and would prevent further damage to biodiveristy and the Earth's ecosystems.

Certain nongovernmental organizations (NGOs) have already begun to draft such a protocol. The drafted protocol would require exporters of biotechnology products to submit complete safety information on a case-by-case basis, establish an indepernent international body of experts to conduct risk assessments and make decisions on all transboundary trade of genetically modified organisms (GMOs), include public participation at every step of decision making, require mandatory labeling for genetically engineered food products provide technical and analytical support for member countries, and establish liability standards. The protocol also calls for risk assessment to include social and cultural studies and states that genetic diversity "is dependent on the socioeconomic conditions of the peoples maintaining it" (Community Nutrition Institute 1996, 30). However, the United States would most likely only adopt a significantly modified position, that is one that is science-based and within a framework of risk assessment and management that has been proven adequate in the United States.

Certain others in the scientific community do not support the adoption of the CBD and its legally binding International Biosafety Protocol and believe that regulation of biotechnlogy would serve as a hazard to the diffusion of biotechnology in the developing world; stifle the development of certain applications of biotechnology such as those that can assist in toxic waste removal, water purification, and displacement of agricultural chemicals; and would no likely meet the protocol's goal of safety enhancement cost-effectively.

CONCLUSION

Techniques of biotechnology and genetic engineering, if applied responsibly, have the potential to increase productivity in crops and livestock, control pests, produce new food and fiber crops, and develop effective medicines. Potential environmental and economic benefits from biotechnology include the reduction of fossil fuel in agriculture and forestry through improved nutrient availability in crops and livestock, use of fewer artificial inputs (e.g., synthetic nitrogen fertilizers, insecticides, and fungicides), and more cost-effective and environmentally friendly waste management practices such as bioremediation. If realized, these improvements will help protect ecological systems by reducing habitat degradation. Although biotechnology and genetic engineering can be expected to provide major benefits to agriculture and the environment, risks with the use of this technology should also be recognized.

Environmental risks include: the potential to alter basic interactions in natural ecosystems; create plants that become new weeds; release pest control organisms that evolve resistance or harm nontarget organisms, or both; and deplete biodiversity further. Potential health risks include: the possibility to introduce new allergens into the food supply increase levels of antibiotic residues in the food supply, increase the prevalence of antibiotic-resistant bacteria, and create new virulent strains of bacteria and viruses. Socioeconomic risks include: the ability of transnational companies (TNCs) to create monopolies and exploit farmers in developing countries. Ethical risks include issues related to the inhumane treatment of animals and conflicts arising in value and belief systems of certain individuals and organizations.

The public appears to find certain applications of biotechnology and genetic engineering acceptable, including plant-to-plant gene transfers and use of this technology to solve food problems in developing countries. However, other applications of biotechnology, such as animal-to-plant gene exchanges are reported to be less acceptable. The majority of consumers across international borders support mandatory labeling of genetically engineered foods. Because the goal of biotechnology is to reduce rather than increase risk in the food supply, a more effective international regulatory policy is needed. Adoption of an International Biosafety Protocol, as commissioned by article 19 of the Convention on Biological Diversity (CBD), would help reduce the new and novel risks associated with the use fo biotechnology and genetic engineering.

2

Bleaching Technologies

Paper and cardboard production has greatly increased in recent years, being one of the sectors of the forest industry with important expectations of expansion. the most important feedstock is generally wood, the main components of which are cellulose (42 % dry organic matter), hemicellulose (27 %) and lignin (25%), and to a lesser extent wood extracts such as tannis and resin acids (6%). Cellulose is a linear polysaccharide consisting of β-D-glucopyranose units, which are linked by (1–4)-glucosidic bonds. Hemicelluloses are composed of different carbohydrate units and are located at the interface of cellulose and lignin. Lignin is a complex three-dimensional polymer consisting of non-repeating phenyl propanoid units linked by carbon–carbon and ether bonds. Lignins have been generally classified into three major groups – softwood lignin, hardwood lignin, and grass lignin – based on their chemical structure of monomer units.

The process of cellulose pulp production requires an initial step of pulping after debarking and chipping of the raw material. The purpose of pulping is to remove lignin in order to facilitate fibre separation and to improve the papermaking properties of the pulp. There are basically two methods: mechanical and chemical pulping, accounting for 33 % and a 67 % respectively of the total production. Most chemical pulping is carried out according to either the Kraft (sulphate) process or the sulphite process. The Kraft process entails treating wood chips at 160–180°C with a liquor containing contains sodium hydroxide and sodium sulphide, which promotes cleavage of the various ether bonds in the lignin. The lignin degradation products so formed are dissoved in the alkaline pulping liquor. Depending on pulping conditions, as much as 90–95 % of the lignin is removed from wood at this stage. After separation of the pulp, the spent liquor is evaporated to a high concentration and then burned to recover energy and inorganic chemicals. Neither the Kraft nor the sulphite process removes all the lignin: about 5– 10 % of the original remains in the pulp, because it cannot be removed by extended pulping without seriously damaging the polysaccharide fraction.

TABLE 2.1 : PREDICTED CONSUMPTION OF PAPER AND CARDBOARD

	Consumption (millions of tonnes)		
	1993	**2010**	**Increment (%)**
North America	92 338	144 101	56.0
Europe	65 373	111 916	71.2
Old URSS	4 323	8 561	98.0
Africa	3 305	8 265	150. 1
Latin America	9 194	17 094	85.9
Asia	74 375	183 298	146.5
Oceania	3 244	5 943	83.2
Total	252 202	479 178	90.0

The pulping process plays a central role in the pollution load and composition of wastewater produced at pulp mills. Mechanical and thermomechanical pulping give high yields and accordingly low pollution loads are produced. Their wastewaters usually contain high amounts of biodegradable matter, such as carbohydrates and organic acids. Semi-chemical (e.g.) neutral sulphite (semi-chemical) and chemo-thermomechanical pulping wastewaters are of intermediate strength and contain higher amounts of lignin. In the chemical pulping, a high-strength wastewater is produced with soluble carbohydrates and organic acids as the main organic components, along with hardly biodegradable wood

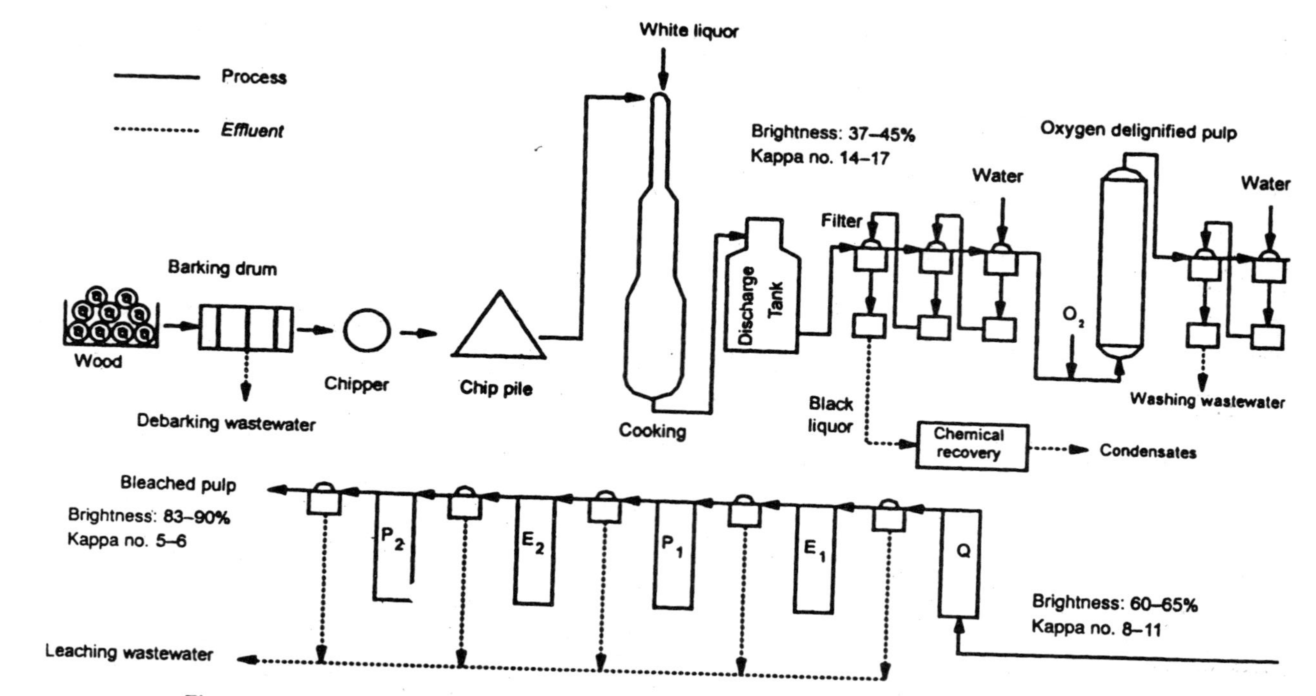

Figure 2.1 : *Flow diagram of Kraft pulp mill including post-digested oxygen delignification and totally chlorine free bleaching process.*

TABLE 2.2 NOMENCLATURE USED FOR THE DIFFERENT BLEACHING STEPS IN THE KRAFT PROCESS

Bleaching processes	**Notation**	**Reaction**
Mill Trials:		
Chlorine	C	Reaction with elemental chlorine in acid solution
Chlorine dioxide	D	Reaction with chlorine doxide in acid solution
Mixture of chlorinated compounds	C_D or D_C	Mixture of chlorine and cholrine dioxide
Sodium hypochlorite	H	Reaction with hypochlorite in alkali solution
Oxygen	O	Reaction with pressurized oxygenin alkali solution
Ozone	Z	Reaction with ozone
Hydrogen peroxide	P	Reaction with hydrogen peroxide
Enzyme (xylanase)	X	Enzymatic hydrolysis of xylan
Alkali extraction	E	Dissolution of reaction products with sodium hydroxide
Monox-L (hypochlorous acid)	M	Reaction with hypochlorous acid
Chelating treatment	Q	Chelating treatment with EDTA
Process Pilot Plant:		
Prenox	NO_X/O_2	Acid pretreatment
Peroxyacetic acid	CH_3CO_3H	Chemical bleaching
Peroxyformic acid	HCO_3H	Chemical bleaching
Dimethyl dioxirane	DMD	Activated oxygen

extracts. Furthermore, inorganic compounds such as sulphates, sulphites, sulphide and alkali are also present at high concentrations.

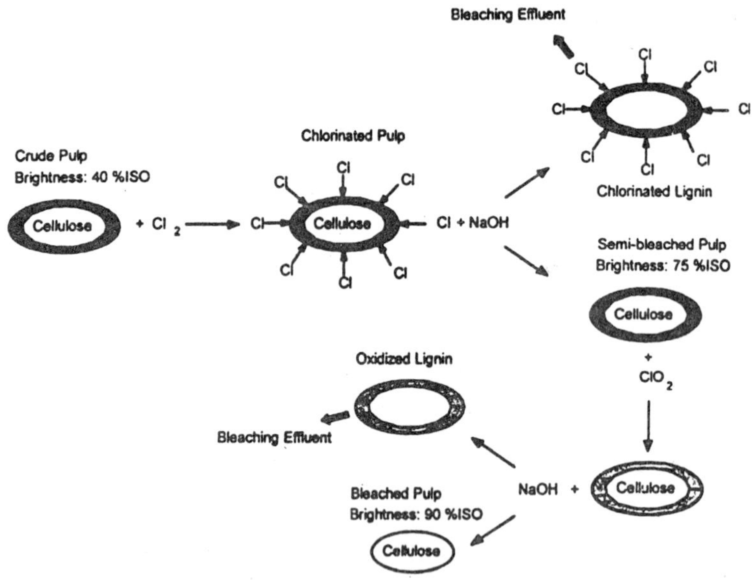

Figure 2.2 : ***Mechanism of conventional bleaching by chlorine and chlorine dioxide.***

Removal of the residual lignin, which is responsible for the dark colour of Kraft pulps, calls for a further multistage bleaching process. The more common nomenclature for expressing the different bleaching stages is shown in Table 2.2. Conventionally, the bleaching of digested pulp by Cl_2 has implied the generation of adsorbable organic halogens (AOX), this being problematic for biological wastewater treatment due to their high toxicity and lwo biodegradability. Moreover, organohalogens have generally been regarded as undesirable xenobiotic compounds. Consequently, due to market pressures and government norms, the pulp and bleaching technologies. New operating policies were developed to minimize the pernicious impact of the free disposal of liquid residues in natural courses and landfills. The proposals towards cleaner processes were focused on bleaching with chlorine-free sequences to eliminate emission of organic halogen compounds.

NEW BLEACHING PROCESSES

The first significant change introduced in the bleaching plant has been the replacement of elemental chlorine by chlorine dioxide; the so-called 'elemental chlorine free' (ECF) bleaching process. This alternative leads to a high-brightness pulp with acceptable strength properties and wastewaters with lower AOX concentrations and lower AOX loads, ranging from 0.7 to 0.9 kg t^{-1} air dried pulp (ADP) for mature eucalyptus and from 0.4 to 1.0 kg t^{-1} ADP for plantation (young) eucalyptus.

Since 1990, 'totally chlorine free' (TCF) bleaching has been used, largely in response to market demands for non-chlorine-bleached pulp. TCF bleaching was made possible after the action of a pre-delignification step with pressurized oxygen, which leads to a pulp with considerably lower kappa number. Then the process can be based on the action of oxygen, ozone, hydrogen peroxide, or enzyme which replaces chlorine and chlorine dioxide bleaching. Peroxide stages are preceded by some treatment, such as an acid wash, and a chelating stage to remove metal ions. Peroxide-based TCF bleaching processes produce virtually no organohalogens as the AOX loads are reported to be below 0.05 kg t^{-1} ADP. In a recent study, analysed the methanogenic toxicity and anaerobic biodegradability of ECF and TCF bleaching effluents from oxygen-delignified eucalyptus Kraft pulp. The effluents from chlorine and ECF bleaching sequences had similar methanogenic toxicities, with 50 % inhibiting concentrations (50 %IC) of 0.65–1.48 g of less toxic, with a 50 %IC of 2.3 g COD/ 1. the fact that the ECF bleaching effluent was no less toxic than that of chlorine bleaching combined with the residual toxicity of TCF, indicates that other substances aside from organohalogens contribute to the high methanogenic toxicity in bleaching effluent.

To obtain an efficient operation both processes, ECF and TCF, have additional requirements such as an improved quality of raw materials, extended cooking, postdigested oxygen delignification, optimized washing to remove high lignin fractions and closure of the brown stock area.

The utilization of biological or enzymatic processes in several areas of food and non food industries has become feasible by the development of highly specialized microorganisms, which can be employed directly or through isolation of the enzymes they produced. It can be expected that progress will be made in reducing chemical inputs by the use of biotechnology, which also provides generally better environment protection. The present status of the possible biotechnical solutions at different process stages in the pulp and paper industry is summarized in Table 2.3. As it can be observed,only few of these solutions are commercial.

TABLE 2.3 :
BIOPROCESSING IN THE PULP AND PAPER INDUSTRY

Process stage and identified problems	Biotechnological solution		
	Identified	Tested	Commercialized
Raw material treatment:			
Debarking	+	+	–
Wood preservation	+	+	–
Mechanical pulping:			
Energ saving	+	+/–	–
Pitch removal	+	+	+
Dissolved colloids	+	+	–
Chemical pulping:			
Delignification	+	–	–
Bleaching:			
Hemicellulase-aided bleaching	+	+	+
Enzymatic delignification	+	+/–	–
Paper manufacture:			
Recycled paper	+	+/–	–
Fibre modification	+	+/–	+/–
Microbial contaminants and slimes	+	+/–	+/–

+/– Results not conclusive

Enzymes (*xylanases*) can used as a biotechonological alternative as a pretreatment; oxidative enzymes from white-rot fungi may be used in the bleaching process itself.

Xylanases are being employed as enzymatic pre-treatments in Kraft pulp bleaching process at mill scale in order to improve delignification by degradation of hemicelluloses. Xylan forms the basic backbone polymers of wood hemicelluloses, depending on the degree of polymerization of the wood species. Elucidation of the exact mechanism of the enzyme-aided bleaching methods is key. There are different hypotheses to explain the enzymatic action on the fibre-bound substrate, but it can be concluded that two types of phenomenon are involved: (1) the hydrolysis of the re-precipitated xylan, formed during delignification, renders the pulp more permeable, thus facilitating the removal of residual lignin; (2) the partial hydrolysis of xylan, located in the inner layers and possible linked to lignin, is likely to facilitate further bleaching. A pretreatment step with xylanases achieves and increase of brightness or a decrease of chemical consumption. Other positive features are the low cost of enzyme and the low investment costs if the enzyme stage is performed in the brownstock storage tower. However, the use of xylanases will always require some further chemical delignification for complete pulp bleaching and consequently will not permit large chemical savings even when the process operates at the higher enzyme dosage.

ENZYMATIC BLEACHING

White-rot fungi are well known for their outstanding ability of depolymerize and mineralize lignin and to degrade high-molecular-weight pollutants. Several strains have been identified that cause extensive delignification or bleaching of unbleached Kraft pulps (UKPs). The first attempt to bleach Kraft pulps with white-rot fungi was made by Kirk and Yang (1979) with *Phanerochaete chrysosporium*, which lowered the lignin content of the pulp but also attacked the cellulose and significantly decreased the pulp strength. Other white-rot fungi have been found to more selectively delignify Kraft pulps; these include *Trametes versicolor*, *Phanerochaete sordida*, and unidentified white-rot fungal strain called IZU-154 and *Bjerkandera* sp. BOS55.

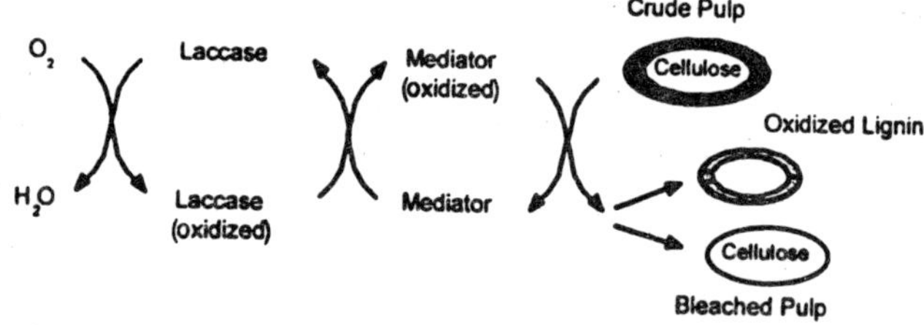

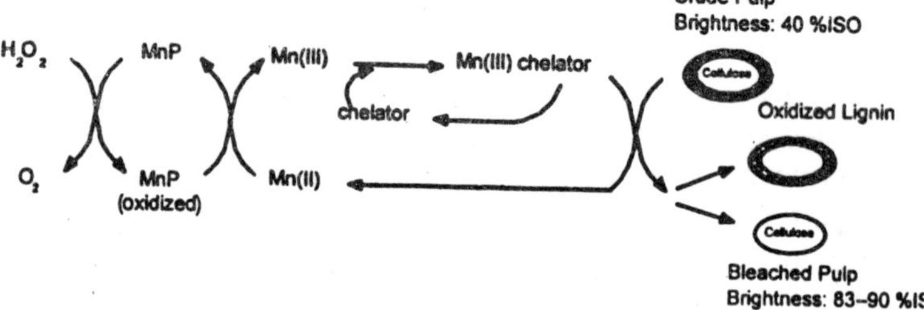

Figure 2.3 : *The oxidative pathway for catalytic action of laccse and manganese peroxidase on lignin.*

Lignin biodegradation is initiated by several extracellular oxidative enzymes excreted by white-rot fungi, including lignin peroxidase (LIP), manganese dependent peroxidase (MnP), manganese-independent peroxidase (MIP), laccase and hydrogen peroxide generating oxidases. Purified ligninolytic enzymes have been shown to cause limited delignification and bleaching of UKP, provided that the hydrogen peroxide is carefully dosed and the enzymes are coincubated with low-molecular-weight cofactors: veratryl alcohol for LIP, manganese, organic acids and surfactants for MnP and *n*-substituted aromatic compounds for laccase. The role of LIP in pulp biobleaching by whole cultures is not clear because this enzyme has generally not been detected during fungal biobleaching in many of good biobleaching strains. Laccase and MnP, on the other hand, are excreted at varying levels by different white-rot fungal cultures when biobleaching occurs.

Laccase is a blue copper phenoloxidase that contains four copper atoms per polypeptide chain and is capable of catalysing the four-electron transfer reaction necessary to fully reduce oxygen to water. Although by definition *p*-diphenols serve as electron donors for laccase, these enzymes have a broad substrate specificity including electron donors for laccase, these enzymes have a broad substrate specificity including polyphenols, methoxy-subtituted monophenols, aromatic amines and a considerable variety of other natural and synthetic substrates. Since *P. chrysosporium*, the model white-rot fungus of choice, was believed not to produce laccase, the role of laccase in lignin degradation has been less intensely studied. Laccase can catalyse the alkyl–phenyl and C_{α}–C_{β} cleavages of phenolic dimers which are used as model lignin substructures and can catalyse the dimethoxylation of several lignin model compounds. The redox potentials of the laccases studied so far have not been thought sufficiently high to remove electrons from the non-phenolic aromatic substrates that must be oxidized, during lignin degradation. Bourbonnais and Paice (1992) showed that an artificial laccase substrate, ABTS (2,2'-azinobis-[3-ethylbenzthiazoline-6-sulphonate]), could act as a redox mediator which enables laccase to oxidize non-phenolic lignin model compounds. The laccase mediator concept combines the action of the enzyme with a low-molecular-weight and environmentally friendly redox mediator, which generates strongly oxidizing compounds that specifically degrade lignin leaving the cellulose fibre intact. Laccase Mediator System technology was tested successfully at a pilot plant trial using pulps of different origins. The performance of this system has been further improved by optimizing the mixture of system components, protecting the enzyme against inactivation and enhancing reaction kinetics to obtain a reduction of the mediator quantity, which is the main operating cost of the process. For many pulps, a 50 % delignification could be obtained using 5 kg of mediator per tonne of pulp. The most efficient mediators found belonged to the *n*-heterocyclic compound group.

Since its discovery in *P. chrysosporium* in 1994, MnP has been found to be secreted by many white-rot fungi. Initially, MnP was thought to play only a limited role in lignin depolymerization, oxidizing only phenolic sub-units, while LIP was regarded as the main enzyme. MnP was found to depolymerize ^{14}C-labelled dehydrogenative polymerizate lignin, to oxidize non-phenolic model compounds, and to discolour dissolved coloured lignin in bleached Kraft mill effluent. As MnP appears to have an important role in lignin biodegradation, factors involved in its catalytic cycle require special attention to determine their relative importance. The requirement of the Mn(II) ion in MnP expression is widely known, and has been reported for a great number of white-rot fungi. Mn(II) is involved in the catalytic cycle of MnP and oxidized to Mn(III). Mn(III) complexed with an organic acid present in the medium is the oxidant capable of diffusing inside the matrix of the polymer and depolymerizing lignin. Accordingly, these complexes are also able to attack a great number of substrates with a phenolic structure similar to that of lignin.

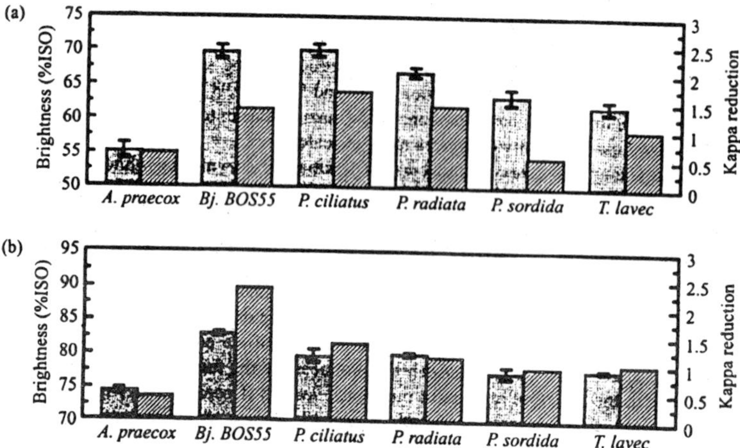

Figure 2.4 : ***Biobleaching and delignification of UKP (a) OKP (b) by selected white--rot fungal strains in a 14-day incubation period after oxalic acid extraction. Strains:*** **Agroybe pracox, Bjerkandera** ***sp. BOS55,*** **Polyporus ciliatus, Phlebia radiata, Phenerochaete sordida** ***and*** **Trametes** ***sp. LAVEC94-3. The initial brightness and kappa number medium, the brightness and kappa umber were 49.0(±0.87); after 14 days of incubation in sterile medium, the brightness and kappa number were 49.0(±0.1) %ISO and 12.94(±1.27), respectively. The initial brightness and kappa number of OKP were 69.4(±0.4) %/ISO and 9.00(±0.20); after 14 days of incubation in sterile medium, the brightness and kappa number were 69.2(±0.3) %ISO and 8.94(±0.27), respectively***

There is increasing evidence that MnP has a key role in the

biological bleaching of pulp, which is supported by several facts. (1) MnP activity is generally detected in active biobleaching cultures of different strains. Moreover, the maximum level of MnP coincides with the time period in which pulp biobleaching occurs. MnP activity has also been correlated with the biobleaching ability of different white rot fungal isolates. (2) Mutants of *Trametes versicolor* deficient in MnP do not bleach, and its bleaching activity is partially restored after addition of MnP. (3) Catalase, an enzyme that destroys hydrogen peroxide, inhibits bleaching. (4) Pulp biobleaching and delignification by purified MnP in cell-free systems were recently demonstrated when Mn(II), Tween 80, malonate and hydrogen peroxide were supplied at appropiate concentrations. Nevertheless, experiments with the purified enzymes as well as *in vitro* incubation of UKP with cell-free extracellular fluids of biobleaching fungal cultures without cofactors have failed to reproduce the extensive biobleachng obtained by whole-culture systems, thus indicating that effective biobleaching is achieved only by the combined action of enzyme, cofactors and stabilizing organic acids. However, the specific contributions of each component to bleaching in whole fungal cultures are still unknown.

In a recent study, showed that *Bjerkandera* sp. BOS55 is able to cause extensive delignification and bleaching of UKP as well as oxygen-delignified Kraft pulp (OKP). The brightness and kappa number reduction achieved for both pulps over a 14-day incubation period by five selected fungal strains after oxalic acid extraction. Brightness gains of up to 20 % ISO in UKP and up to 14 % ISO units in OKP were obtained, giving high final brightness values up to 68 nd 80 % ISO units, respectively. The residual lignin in OKP is therefore less bleachable by various white-rot fungal stains, which is logical because residual lignin in UKP and OKP differ in both amount and structure, as oxygen delignification removes about one half of the residual lignin in Kraft pulp and produces lignin modifications yielding highly condensed lignin units. The specific kappa number decrease per unit of brightness gain during the biobleaching of OKP was as high as 0.20 with the best strain, *Bjerkandera* sp. BOS55. During the course of biobleaching and delignification, MnP was the major oxidative enzyme activity detected in the extracellular peroxidases under a wide variety of culture conditions. Therefore it seems that other components of the bleaching system, such as secondary metabolites, organic acids and hydrogen peroxide, should be considered as possible rate-limiting factors. Surprisingly, *Bjerkandera* sp. BOS55 also presented a manganese-independent bleaching system under manganese-deficient conditions, MnP being the predominant oxidative enzyme produced. Consequently, biobleaching by *Bjerkandera* sp. BOS55 in the absence of manganese must be attributed to other peroxidases or to a presently unknown cofactor for MnP other than manganese. The incomplete definition of the mechanisms which explain the bleaching process is an important obstacle for the complementation of this new process at an industrial scale. Likewise availability of enzymes; neither enzyme is currently available in sufficient quantities for mill trials, and scale-up of enzyme production from fungal cultures may be costly.

PRODUCTION OF MANGANESE-DEPENDENT PEROXIDASE

Because the synthesis of ligninolytic enzymes occurs during secondary metabolism in response to nitrogen or carbon limitation the productivity of the process is quite low, which constitutes on of the major obstacles to implementing bleaching sequences by using oxidative enzymes at large scale.

Several attempts to achieve production of ligninolytic enzymes by white-rot fungi in submerged or immobilized liquid cultures have been described. Most of them are completely mixed systems, with or without mechanical agitation, such as completely stirred tank reactors and airlift reactors or bubble columns. Only relatively sharp transient peaks of activity and a later decrease of its maximum level in successive batches have been observed in most cases, regardless of the strain or the carbon source used. One of the factors probably involved in the destabilization of the enzymatic complex is the action of extracellular proteases which are produced by the fungi in nutrient starvation conditions in the search of alternative nutritional sources. The enzymes produced are quickly denatured by the action of these proteases and the final attack on the cell membrane as an ultimate nutrient source means lysis of the culture. From the analysis of results of batch experiments, other environmental factors (a high oxygen tension, gentle agitation, presence of different cofactors such as Mn(II) for MnP, veratryl alcohol for LIP or *n*-substituted aromatic compounds for laccase) are necessary to achieve maximum enzymatic titres in the cultures.

When operating with pellets as well as with immobilized mycelia, the branching of the hyphae favours linking of pellets or immobilized bioparticles into conglomerates, which provokes many operational problems such as fouling of the probes, blockage in the sampling and feeding lines and broth viscosity increase. Moreover, the operation with nitrogen-limited medium enhances the production of extracellular polysaccharides, which act as linking agents. All these factors cause limitations in the mass and oxygen transfer coefficients and negatively affect the overall efficiency. These drawbacks make it necessary to control mycelial growth in processes operated for long periods of time, an objective that is an important goal in the proper performance of any type of bioreactors.

Application of a perturbation in the form of gas pulsation can minimize the problems related to excessive growth of mycelia and thus avoid clogging. This pulsation is provoked by means of a pulsing device, which can be regulated in amplitude and frequency. This pulsation systme increases the overall productivity of different biotechnological processes.

The effect of the pulsation of oxygen on MnP production by *Phanerochaete chrysosporium* in two types of bioreactors (fluidized-bed with pellets and packed-bed with immobilized mycelia) has been analysed. In both cases pulsation was a key factor and allowed a stable performance to be achieved. The pulsation modified the morphology of mycelial pellets because the applied shear stress on the bioparticle surface caused by the pulsation limited hyphal extension. In the case of immobilized mycelia the pulsation avoided interconnection between foam blocks and aggregation of conglomerates.

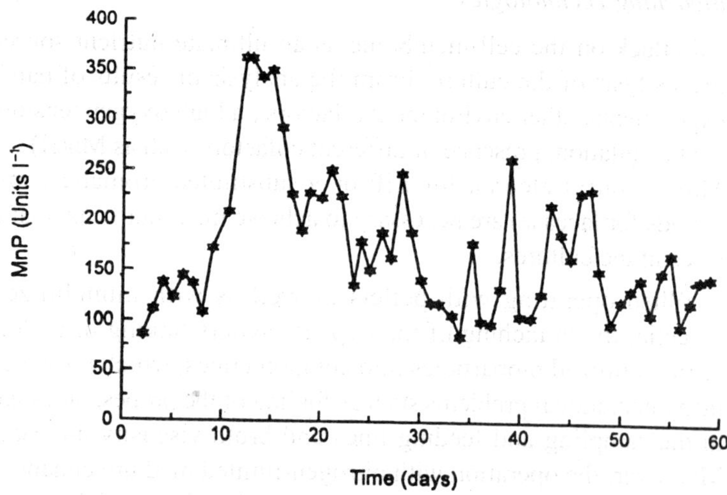

Figure 2.5 ***Continuous production of MnP by immobilized* P. chrysosporium *in a packed-bed bioreactor operated with pulsation at a hydraulic retention time of 24 h and a Mn^{2+} concentration of 5 mM***

The efficiency of the process largely depends on a number of environmental and operating conditions. Optimal conditions for enhanced MnP production in a fixed-bed reactor for an extended

TABLE 2.4 : COMPARISON BETWEEN DIFFERENT SYSTEMS FOR THE PRODUCTION OF MNP IN BIOREACTORS

Reference	Bioreactor/ process	Operation time (days)	Max. activity (Units l^{-1})	Producti- -vity (Units l^{-1})
Bonnarme and Jeffries (1990)	Airlift/batch	3.9	192[a]	49
Bonnarme *et al.* (1993)	Airlift/batch	3.8	365[a]	95
Laugero *et al.* (1996)	Bubble- column/batch	4	726[a]	181
Moreira *et al.* (1997b)	Packed bed/ continuous	140	250	202

[a] The extinction coefficient for the dimeric dimethoxyphinol oxidation product was corrected to 49 600 M^{-1} cm^{-1}.

period of time correspond to a feed rate of nutrients of 250 mg glucose l^{-1} h^{-1} and 0.6 mg ammonia l^{-1} h^{-1}, a Mn(II) concentration of 5 mM, adequate oxygen supply, a bioreactor hydraulics of a plug flow with partial mixing and an operating hydraulic retention time of 24 h. Table 2.4 shows that the results obtained following this strategy compare favourably with the best recently reported for the production of MnP, even when mutant cells of *P. chrysosporium*, strain INA-12 were used.

FUTURE PERSPECTIVES

The application of ezymatic bleaching to the pulp and paper industry is still to be developed. The presently commercial xylanases reduce the need to bleach chemicals for high-brightness pulp, and they have relatively low cost and ease of application with great selectivity. The drawback of the hemicellulase-aided bleaching is that it is an indirect method, not directly delignifying pulp. Both laccase and MnP can achieve more substantial delignifying action than xylanase, but there are obstacles to be overcome before either enzyme can be used in a cost-effective manner.

To make the application of ligninolytic fungi and their oxidative enzymes feasible in a biobleaching process serious reduction in the time from days to hours is needed, which is only possible through knowledge of the mechanism involved. For laccase, so far no mediator is sufficiently effective and inexpensive to be commercially viable. There is currently no large scale commercial source for enzyme production, so the costs of the process associated with the use of oxidative enzymes can not be accurately caluclated. Future research should allow for a more efficient utilization of the bleaching system, friendly clean process of low cost. Thus, oxidative enzymes, which can be regarded as catalysts for oxygen and hydrogen peroxide-driven delignification, may also find a place in the pulp bleaching process in coming years.

3

Genomic Repression

Specific base-pairing in nucleic acids is fundamental to the storgae and retrieval of all genetic information. During the last two decades, concerted efforts have been made to harness this simple and universal process for the control of aberrant gene expression. These efforts have been directed at three different levels: (1) interference with the regulatory regions of specific genes by means of short, complementary single-stranded oligonucleotides that can form a site- specific triple helix and thereby interfere with the binding of regulatory proteins at the gene promoter; (2) inhibition of the messenger ribonucleic acid (mRNA) function by the formation of double-stranded nucleic acids with antisenre oligonucleotides; and (3) sequence-specific degradation of the RNA molecules by catalytic RNAs known as ribozymes. Depending on the nature of the pathological condition, each of these approaches may present certain merits and disadvantages. The purposes of this chapter are to extend these discussions and to provide an overview of the current status of ribozyme and triplex deoxyribonucleic acid (DNA) technology. Furthermore, we will attempt to point out the therapeutic potential of the oligonucleotide-based drugs against infectious agents and degenerative disorders

Nucleic acids are composed of polymeric chains of adenine and guanine (purines), thymine/uracyl or cytosine (pyrimidines), and monophosphate residues covalently linked through phosphodiester bonds. Short strands of nucleic acids (10–40 bases) are generally called oligonucleotides. Two complementary strands of nucleic acids can form a noncovalent duplex as a result of Waston–Crick base pairing, for adenine can form a hydrogen bond with thymine or uracyl, whereas cytosine can bond with guanine. In the case of DNA, where only the 2-deoxy derivative of the ribose is utilized, one strand serves to store the genetic code and is called the sense strand, whereas the other provides the complementary supporting strand and is known as the antisense strand. Ribonucleic acid is generally copied from the antisense strand and has the same sequence as the DNA sense strand except that it uses the unmodified ribose and also the base uracyl in the place of thymine. Additionally, a third strand of nucleic acid can bind to the DNA duplex through a non-Watson–Crick (Hoogsteen) type of hydrogen bonding, resulting in the triple helical or *triplex* DNA structure. The ability of the nucleic acids to form both the duplex and triplex structures is presently being utilized to develop therapeutic agents for inhibiting specific gene function.

INHIBITION OF SPECIFIC GENE EXPRESSION BY TRIPLEX-FORMING OLIGONUCLEOTIDES

Several investigators have proposed the possibility of triplex DNA formation resulting from intramolecular rearrangement of the DNA double helix within the natural genome. Immunofluorescent staining of fixed metaphase chromosomes with the triplex-specific monoclonal antibodies, Jel 318 and Jel 466. provided more direct evidence for the existence of such triplex structures in normal cells. Thomas et al. reported that hydralazine (an antihypertensive drug) elicits antinuclear antibodies owing to its ability to stabilize triplex structures. Serum samples derived from hydrazine-treated patients showed the presence of a high-affinity antibody directed to the triplex DNA. Furthermore, DNA binding activity of these antibodies was found to be significantly lower in the absence of any triplex-stabilizing agent. These data suggested that a possible mechanism for antinuclear antibody production in hydralazine-treated patients might involve the induction and stabilization of immunogenic forms of DNA, including higher-order structures such as triplex DNA.

Unlike the intramolecular triplex due to the rearrangement of the duplex DNA, the intermolecular DNA triple helix formation involves the binding of a separate third intermolecular DNA triple helix formation involves the binding of a separate third strand that offers the opportunity to achieve precise sequence recognition and can, therefore, be harnessed for gene therapy. The formation of a specific complex between two strands of polyuridilic acid with one strand of polyadenylic acid in the presence of divalent cations was first reported in 1957. More recent studies have revealed the site-specific formation of short intermolecular triplexes by a wide variety of detection techniques including footprinting, affinity cleavage, and nuclear magnetic resonance (NMR). Chemical and physical evidence indicates that the nucleotide bases of the third strand occupy the major groove of the Watson–Crick double helix. The definition of a triple helix in nucleic acid results from specific Hoogsteen-or reverse Hoogsteen-type hydrogen-bonding interactions between the bases in a homopurine strand of a Watson–Crick duplex and an additional third strand. Several DNA triplets have been characterized that fall into two distinct classes of the structural motifs. In the case of pyrimidine–purine–pyrimidine (YRY, indicating Hoogsteen bond) triplexes (so-called "pyrimidine motif"), the third pyrimidine-rich strand runs parallel to the purine strand of the target duplex. It results in the formation of TAT and $C^{+}GC$ combinations. The requirement for protonation of cytosines in the third strand suggests that these triplexes are stable only at low pH values (<6.0). In the second class RRY (purine–purine–pyrimidine triplex), the third purine-rich strand runs antiparallel to the duplex DNA and is generally purine rich. The best-characterized triplets within this "purine motif" are AAT and GGC. The latter types of triplexes are stable at physiologic pH values and at 37°C. Recent studies have also revealed the possible existence of several new triplet combinations: GTA and TCG within YRY class of triplexes and a protonated triplet $A^{+}GC$ within RRY triplexes. These noncanonical triplets allow the extension of the existing triplex possibilities to all four naturally occurring base pairs in the target duplex.

Binding of these triplex forming oligonucleotides is sequence specific. Once a triple-helix-forming oligonucleotide is attached to the target sequence, it can alter the affinity of DNA-binding proteins like restriction endonucleases, DNA methylating enzymes, transcription factors, and both DNA and RNA polymerases. Wrange et al. have shown that triple helix DNA alters nucleosomal protein–DNA interactions and may act as a nucleosome barrier. Helene and coauthors have found that triple-helix-forming oligonucleotides inhibited transcriptional elongation. Therefore, triplex-forming oligonucleotides are considered to offer significant potential for modulating gene expression with novel stategies for gene therapy.

Triple-helix-forming oligonuleotides are already being explored as promising antiviral and anticancer agents. Different sequences withing HIV-1 and HIV-2 genomes were successfully targeted to modulate various steps of the infection process, including the initiation of viral transcription, transcription elongation, reverse transcription, and integration. In recent years, several laboratories have attempted to inhibit the expression of oncogenes with triplex forming oligonucleotides. In an effort to develop specific transcriptional inhibitors of the human Ki-*ras* oncogene, Mayfield et al. designed oligonucleotides targeted to the 22 bp pyrimidine–purine motif in the human Ke-*ras* promoter (–126 to –147). These authors have shown that oligonucleotide-directed triplex formation inhibits sequence specific nuclear protein binding to the Ki-*ras* promoter. Similar results were obtained by Miller et al. and Thomas et al. when they used sequences from the upstream region of the oncogene c-*myc* as the target. Olivas and Maher were able to suppress transcription of the p53 antioncogene using sequence-specific triple-forming oligonucleotides. Furthermore, a very stable and specific triplex with the promoter of the c-*pim*-1 proto-oncogene was found to block c-*pim*-1 promoter activity in the cell culture system. These data suggest that triplex formation by the oligonucleotides may provide a convenient approach to inhibit transcription of certain proto-oncogenes specifically. Additionally, steroid receptors have been targeted by triplex-forming oligonucleotides. Ing et al. have successflly used a 38-base, single-stranded DNA that forms a triple helix on progesterone response elements that block the progesterone receptor interaction with its target gene. The 38-base, single stranded DNA also inhibited progesterone receptor-dependent transcription in vitro. Furthermore, the cholesterol derivative of the same oligonucleotide specifically inhibited progesterone receptor-dependednt transcription *in vivo*.

Triplex structures may also be involved in the physiological control of specific gene expression. On the basis of results of single-strand specific nuclease accessibility and DNase I footprinting, we have proposed a dynamic model for transcriptional control of the androgen receptor (AR) gene through intramolecular triple helix at the homopurine– homopyrimidine (pur-pyr) element of the promoter region. The model proposes that the pur-pyr region of the AR promoter exists in a conformational equilibrium between a B double-stranded DNA or "B form" and two "H forms" involving intramolecular triple helices (Fig.). In contrast to (CGC) pyr-pur-pyr, the pur-pur-pyr (GGC) structure can form a stable Hoogsteen hydrogen bond at the physiological pH and therefore is considered to be the preferred conformation. The two DNA-binding proteins (Sp1 and ssPyrBF) that specifically bind to this element selectively interact with either one of the B or H conformations. When the pur-pyr element is in the double-stranded B conformation, Sp1 can interact and accumulate at the pur–pyr site, resulting in transcriptional activation. However, when the element is in a pur–pur–pyr H form structure, interaction with the single-strand pyrimidine specific binding factor (ssPyrBF) can stabilize the triple-helical conformation, thereby preventing the access of Sp1 and inhibiting the gene transcription. Varying ratios of ssPyrBF to Sp1 could potentially play a critical role in the spatio–temporal regulation of the AR gene expression.

By coupling the oligonucleotide to various DNA damaging agents, a covalent modification of the target duplex can be achieved. Such modification is found to occur in a site-specific manner. Several approa1ches, including DNA cleavage by EDTA-Fe(II), crosslinking of two strands by psoralen and chlorambucil,

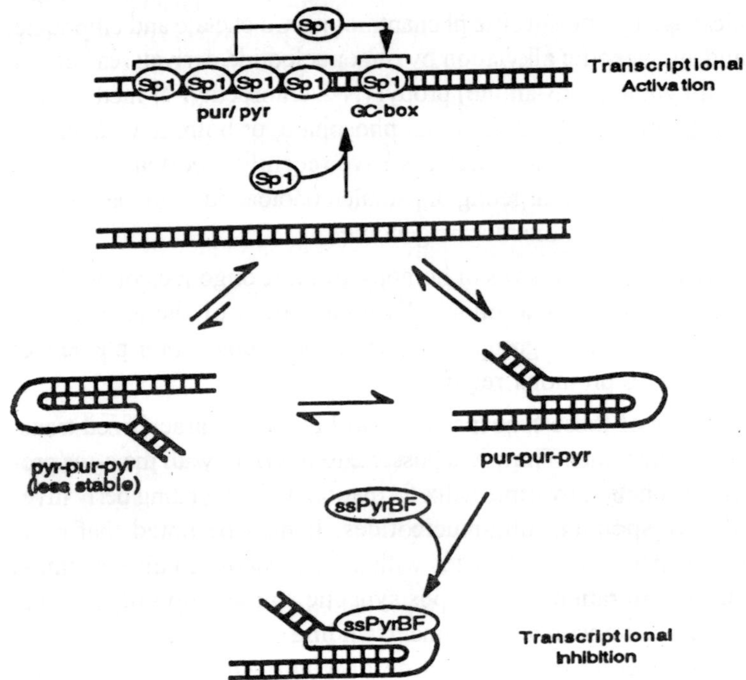

Figure 3.1 : ***A model for transcriptional control at the AR homopurine–homopyrimidine element by intramolecular triple helix.***

The essence of the model is that the conformational structure of the pur–pyr element can alternate between a B-DNA double-stranded form and two H forms involving intramolecular triple helices. In contrast to (C*GC) pyr–pyr, the pur–pur–pyr, the pur–pur–pyr (G*GC) structure can form a stable Hoogsteen hydrogen bond (*) at the physiological pH and therfore is the preferred conformation. The two DNA-binding proteings (Sp1 and ssPyrBF) specific to this element can only bind to particular structures. When the pur–pyr element is in double-stranded conformation, Sp1 can interact and accumulate at the pur–pyr element is in double-stranded conformation, Sp1 can interact and accumulate at the pur–pyr site and provide a readily available source for the GC box located immediately downstream. This situation will enhance transcription. When the element is in a pur–pur–pyr H form structure, the binding of the single strand pyrimidine-specific factor (ssPyrBf) can stabilize the triple-helical conformation, there by preventing the binding of Sp1 and removing the nearby supply source of this transcription factory for the fuctional GC. Varying ratios of ssPyrBF to Sp1 could box potentially play a role in the differential regulation of the AR gene.

cleavage by site-specific phenanthroline-Cu chelate and ellipticine, and site-specific alkylation by oligonucleotide derivatives bearing an alkylating 4-(3-amino) propyl (N-2-chloroethyl-N-methyl) aniline group at 5' or 3' terminal phosphate, or both, have been described. Glazer and coauthors have recently reported successful triplex-mediated targeting of psoralen photoadducts in mammalian cells and within the murine genome.

Reactive derivatives of homopyrimidine oligonucleotides bearing the 5'- or 3'- terminal DNA alkylation of the aromatic 2-chloroethylamino group were used to form covalent triple helices at the c-*fos* promoter region.

Some of these sequences fall within well-characterized trans-acting elements. Thus, it is possible to interfere with the c-*fos* promoter function by triple helix formation with alkylating derivatives of corresponding oligonucleotides. It may be noted that c-*fos* proto-oncogene is associated with the control of cell differentiation and proliferation and the postsynaptic transduction of neuronal signals. To obtain a better understanding of the potential of triple-helix-forming oligonucleotide alkylating derivatives as site-specific transcription modulators of transcription, we examined the cellular system designed for measuring in vivo effects of these oligonucleotide reagents on transcription. A reporter gene strategy coupled with the DNA affinity modification technique was used in an effort to facilitate oligonucleotide uptake, optimize triple helix formation, and provide convenient biological assays. Several key features of the system included facilitation of triple helix formation between the third-strand oligonucleotide bearing the reactive group on its terminal phosphate and a promoter region of the reporter *fos*-chloramphenicol acetyl transferase (CAT) plasmid DNA before transfection. Transfection of the complexes formed *in vitro* followed the measurement of resulting transciptional effects and demonstrated that four different DNA targets within the c-*fos* promoter region could form triplex structures with synthetic oligonucleotides in a sequence-region could form triplex structures with synthetic oligonucleotides in a sequence-specific manner. Moreover, modifications of the c-*fos* promoter at position –83 in front of the cAMP-responsive element and FBS3/AP-2 like site at position –431 by triple-helix-forming oligonucleotides cause dramatic suppression of the CAT activity in endothelial cells.

Although the formation of intermolecular DNA triple-helices offers the potential for designing therapeutic compounds with exquisite sequence-recognition properties, the binding of the third-strand oligonucleotide is certainly less stable than the sense and antisense duplex. One way of improving the interaction is to design compounds that bind to the triplex, but not to the duplex, DNA. There are three different grooves in triple-helical DNA that can be targeted by ligands that bind either covalently or noncovalently to the base pair edges. Because the third strand aligns with the major groove of the Watson–Crick duplex, the unoccupied minor groove has been the principal target for binding ligands that had previously been targeted to the same groove in duplex DNA. In recent years, several such compounds have been described, including 2,6-disubstituted amidoanthraquinones, the benzopyridoindole derivatives, BePI and BgPI coralyne, naphthoquinoline derivatives, and imidazothioxanthones.

In recent years, many laboratories have extended the triplex strategy to RNA targets with secondary structures. It was recently reported that RNA strands are excluded from triplexes containing the purine motif. On the other hand, studies with sequences corresponding to the various combinations of DNA and RNA strands show that two types of triplexes, namely DNA:RNA DNA and RNA:RNA DNA, are unstable with an RNA purine strand. The formation of double-hairpin complexes with an RNA stem-loop target and an antisense oligodeoxynucleotide generates these unfavourable triplexes. Pascolo and Toulme (1996) have demonstrated that the double-hairpin complex strategy allowed formation of a triple-stranded structure with an RNA second strand. This approach offers the possibility of blocking RNA function by selectively targeting the RNA hairpins.

mRNA DEGRADATION BY SEQUENCE-SPECIFIC RIBOZYMES

Regulation of gene expression can occur at several levels, including transcription, mRNA stabilization, and translation. One oligonucleotide-based approach to inhibit expression of secific genes is to target the corresponding mRNA for enzymatic degradation. Once DNA is transcribed into mRNA, a series of mRNA processing events occur (i.e., splicing, polyadenylation, etc.) to produce mature mRNA. Translation of the mRNA can be blocked by either antisense oligonucleotides or by RNA enzymes (ribozymes). Such targeting of the messenger RNA is also a part of the physiological regulatory processes, as indicated by studies with silk moths. Circadian rhythms in the insects are regulated by the Period (Per) gene, which is essential for adult eclosion. The Per gene in the silk moth is expressed in an oscillatory manner and functions as a circadian clock. Although the exact structure of the Per gene in the silk moth is not known, an upstream mechanism regulates the circadian clock in the silk moth by an antisense transcript that can selectively hybridize with the sense messenger RNA. When the antisense RNA is paired with its complementary message RNA molecules, it interferes with the ability of the organism's translation machinery to produce the specific protein. Through this mechanism, the oscillating levels of the Per protein are regulated.

In many cases, the RNA splicing reactions are catalyzed by ribozymes. The RNA enzymes include group I and group II introns, the RNA subunit of RNaseP, hairpin ribozyme, hepatitis delta virus ribozymes, ribosomal RNA, and hammerhead ribozymes. The group I intron was first found in *Tetrahymena thermophia* by Cech and coworkers. The natural function of the group I ribozyme is to process ribosomal RNA. Splicing of the rRNA intron involves a two-step mechanism. In the first step, guanine nucleotide is added to the 5' end of the intron as the intron–exon junction is cleaved. In the second step, the freed 3' exon attacks at the 3' intron–exon junction to release the intron and produces a spliced exon. Ribonuclease P was initially identified in bacteria by Altman and coworkers. The ribozyme consists of a small subunit protein and a catalytic RNA. The biological role of the ribozyme is to generate the mature tRNA by endonuclecatalytic cleavage of precursor tRNA. The hepatitis delta virus is a helper virus, and its RNA was observed by Wu and coworkers. The hepatitis delta ribozyme has autocatalytic RNA processing activity. The hammerhead ribozyme, discovered in viroids by Symons and coworkers, is synthesized in a rolling circle and cleaves the polycisteronic message into individual RNA transcripts. The hairpin ribozyme, described by Bruening and coworkers, is found in plant viroids and has a similar biological function as the hammerhead ribozyme.

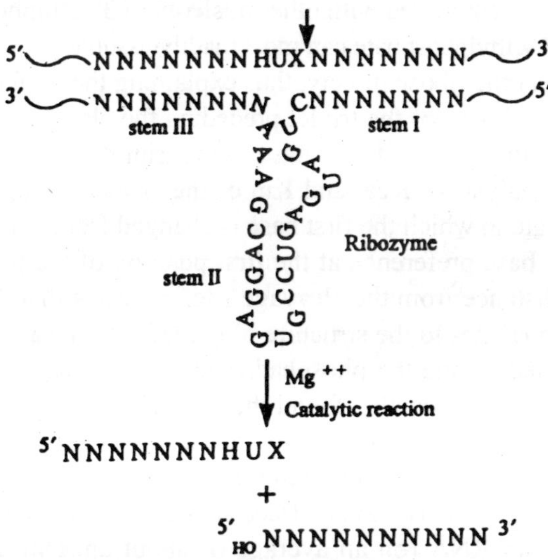

Figure 3.2 : ***Reaction scheme for*** **trans-*hammerhead ribozyme. The hammerhead ribozyme binds to its target RNA substrate throgh stems I and III of the hammerhead ribozyme, which hybridizes with the substrate RNA (N:N'). Stem II is part of the catalytic core. The arrow represents the cleavage site of the substrate (H can bge any nucleotide; X is A, C, or U). Cleavage of a phosphodiester bond occurs at a unique site in this motif, gnerating 2'3'-cyclic phosphate and 5'-hydroxyl terminus.***

The hammerhead ribozyme is one of the smallest RNA enzymes. Naturally, is acts "in *cis*" during viral replication by the rolling circle mechanism. The hammerhead ribozyme has been developed by Uhlenbeck "in *trans* " against other RNA molecules. The *trans*-acting hammerhead ribozyme consists of antisense sections (stems I and III) and a catalytic domain with a flanking stem II as well as a loop section. The mRNA substrate and the hammerhead ribozyme hybridize through stem I and stem III, orienting the-HUX- nucleotides (H can be any nucleotide; X can be an adenine, or a cytosine or uridine, but not a guanine). The GUC nucleotide triplet yields a higher activity of cleavage when compared with other cleavage triplet sequences such as GUA, CUC. Because of this target specificity, the RNA phosphodiester bond immediately following the nucleotide residue, which has its ribose sugar group held in a south conformation (that is C2–endo-C3' -Exo) is most preferably cleaved by the hammerhead ribozyme. Furthermore, compared with other nucleotidyl 3' -ethlphosphates, cytidine 3' -ethylphosphate can most readily assume the south conformation at the ribose moiety, thus explaining the preference for C at the third base of the triplet preceding the cleavage site. The guanine is the preferred first base in the triplet. This is deduced from the analysis of Kcat and Km of the cleavage reactions using substrate in which the first base is changed from G to another base. The base preference at the first position of the triplet, despite its distance from the cleavage site, indicates that the entire triplet contributes to the structure of the transition state intermediate formed during the phosphadiester bond cleavage reaction. However, it has been also found that the optimal triplet depends upon the binding arm composition of the stem I and stem III and the nucleotide lengths of the stems as well as secondary structure of the hammerhead ribozyme. Because -UX- dinucleotides occur frequently in RNA (on an average of about once in every six nucleotides), the hammerhead ribozyme theoreticallty can be designed to cleave any targeted mRNA. Cleavage specificity and efficiency of the hammerhead ribozyme are determined by the binding arms (stems I and III) that hybridize to the complementary sequences flanking the -UX-cleavage site within the targeted mRNA. If stems I and III, which are formed by the hammerhead ribozyme and its substrate, are short enough, the hammerhead ribozyme easily dissociates from the cleavage products, allowing the hammerhead ribozyme to bind to another substrate molecule. By repeating these cycles, one molecule of the hammerhead ribozyme can catalyze several substrate molecules in a mulitple turnover reaction. The cleavage reaction catalyzed by the hammerhead ribozyme follows the *Michaelis–Menten Kinetics*. Kinetic studies of numerous cleavage reactions have demonstrated that the binding of the substrate RNA to the hammerhead ribozyme is equivalent to the rate of helix formation of RNA duplexes. It has been known that the length and sequence of stem I and stem III affect the dissociation between the hammerhead ribozyme and the mRNA substrate as well as the catalytic properties of the hammerhead ribozyme. Theoretically, the arm lengths of each section of stems I and III are optimized at 10–20 nucleotides. However, optimization should also be considered with other factors such as the secondary structure of the RNA substrate and avoidance of a U–G wobble pair by creating A-rich sequences. Computer analysis of the RNA secondary structures based on the energy minimization method can predict primary and secondary stuctures of the targeted RNA substrate for open loops that contain the potential cleavage site for hammerhead ribozyme. Additionally, a comprehensive and elegant procedure for the selection of RNA enzymes, referred as systematic evolution of ligand by exponential enrichment, has been developed and successfully used. The principle is to synthesize a large pool of random-sequence RNA molecules using a bacteriophage RNA polymerase. This pool is subjected to a selection procedure in which the molecules with the desired chemical or physical properties are separated from the rest of the pool. Using reverse transcription and polymerase chain reaction (PCR), a DNA pool is then generated from which, in turn, a new RNA pool, enriched with the desired molecules, can be made. This method was successfully used in the selection of the hammerhead ribozyme capable of cleaving the growth hormone RNA substrate.

The cleavage of the RNA substrate by the hammerhead ribozyme occurs via a transesterification reaction which generates 5' hydroxyl and 2'-3' cyclic phosphate termini. Divalent cations are required for catalysis and enhance the structural stability of fold RNA. The hammerhead ribozyme is one of the smallest RNA enzymes known and may have a potential in specific gene therapy. It has been shown that the inhibition of gene function by the hammerhead ribozyme is primarily due to the catalytic action rather than to an antisense effect. However, the antisense effect also contributes to suppressive function of the hammerhead ribozyme. Initial studies by Sarver *et al.* showed that the hammerhead ribozyme inhibits the replication of HIV-1 in culture cells. Since then, other investigators have reported that the hammerhead and other types of ribozymes inhibit different viruses in cell systems, including the HIV. Application of this approach to other fields such as inhibition of cancer cell growth has also been investigated. The hammerhead ribozyme has been extensively studied, not only with respect to its mechanisms of its action but also for its possible application in vivo. For example, a hammerhead ribozyme designed to cleave mRNA encoding an activated *ras* oncogene in a tumor cell line resulted in a reversion of the malignant phenotype. In this case, the hammerhead ribozyme selectively cleaved the activated *ras* sequence but did not cleave the normal *ras* sequence.

A ribozyme-expressing transgenic mice model has also been generated. Efrat *et al.* found a 70% reduction in the level of glucokinase activity under the control of rat insulin II gene promoter compared with nonribozyme transgenic mice. Similarly, Larsson and coworkers observed a 70, 22, and an 81% reduction in β-microglobulin mRNA in three different transgenic mouse lines expressing the hammerhead ribozyme under the control MMTV-LTR promoter. Another group reported a 78, 58 and 50% reduction of bovine α– lactoglobulin mRNA in three different transgenic mouse lines expressing a hammerhead ribozyme under the control of the CMY promoter. All of these studies indicate the potential of using the ribozyme transgene in experimental animal models for human gene therapy.

For cell culture studies, two major approaches are generally used to introduce hammerhead ribozyme into the cell system. Each has its own advantages and limitations. The first strategy involves synthesis of the hammerhead ribozyme gene *in vitro* and incorporation of the *in vitro* synthesized ribozyme into the cellular environment. The second approach involves subcloning the hammerhead ribozyme gene into a mammalian expression vector and transfection of the expression vector to produce the hammerhead ribozyme in cellulo. A number of transfection methods are used to introduce hammer head ribozymes into cells. These include the N-[1-(2,3-dioleyloxy)propy l]-N,N,N,- trimethylammonium chloride (DPTMA) cationic liposome-mediated transfection, electroporation, microinjection, or calcium phosphate coprecipitation. Malone et al. reported that a few micrograms of synthetic RNA are associated with ten million cells by the DOTMA technique. This corresponds to a delivery of approximately one million molecules per cell. Microinjection is a highly efficient method to deliver the hammerhead ribozyme into cells and can be useful to test the intracellular behaviour of hammerhead ribozymes. The natural hammerhead ribozyme is prone to degradation by nucleases present in both the intra-and extracellular environments. A modified hammerhead ribozyme resistant to nucleases can be designed to increase its rate of accumulation in the cell, thereby decreasing the doses required for effective inhibition of the target gene. Recently, different modified ribozymes have been generated. A DNA–RNA chimeric ribozyme in which stems I and III of the hammerhead ribozyme are replaced by DNA increases its stability *in vivo*. Santora and Joyce used an *in vitro* selection procedue to develop a DNA hammerhead ribozyme that can cleave almost all targeted RNA substrate under physiological condition. The DNA enzyme has a high efficiency on the targeted RNA molecules. The development of a synthetic ribozyme resistance to nuclease degradation is a key consideration for its successful use in pharmaceutical technology. However, for a prolonged inhibition, the use of ribozyme cloned into a mammalian expression vector appears a preferable approach. Two ribozyme-expression strategies have so far been tested with success. First, the hammerhead ribozyme gene is inserted behind a stong promoter for RNA polymerase II, which may be of viral origin, a retroviral long-terminal repeat, or a strong endogenous promoter such as actin gene promoter. A polyadenylation signal is added at the 3' end of the gene to enable transcription termination and the addition of the poly (A) tail which, together with the m7G cap, can increase RNA stability and transport to the cytoplasm if required. An intron can also be inserted to ensure transcription through the normal splicing machinery in the nucleus, which may provide an enhancement of the hammerhead ribozyme effect. Promoters that are transcribed by RNA polymerase II are generally not suitable for production of short RNA molecules such as the hammerhead ribozyme. In this system, at least several hundred nucleotides are needed between the promoter and the terminator for effective transcription. Although extra sequences can be added at both ends of the hammerhead ribozyme, the additional sequences can have undesirable effects on cleavage activity. Also, a hundred-or thousandfold molar excess of hammerhead ribozyme expression over the target RNA is generally required for successful suppression of gene expression. However, one of the main advantages of the RNA polymerase II promoter is the availability of a tissue-specific and regulatable promoter function. Such specificity is essential when hammerhead ribozyme expression is desired only in certain tissues or when expression needs to be turned on or off within a given tissue with regulatable switches.

The hammerhead ribozyme transcripts usually accumulate to relatively low levels and only within the cytoplasm. An alternative to RNA polymerase II transcription is to use an RNA polymerase III specific promoter. The RNA polymerase III transcribes a variety of small nuclear and cytoplasmic RNAs that are abundant in all cell types. The RNA polymerase III-specific promoters from genes encoding tRNA, U6 small nuclear RNAs, and "virus-associated" RNAs that are expressed at high levels in adenovirus-infected cells, have been used to drive the expression of hammerhead ribozymes. The hammerhead ribozyme under the control of the RNA polymerase III promoter is expressed at high levels in the cell culture system. Cotten and Birnstiel observed that the ribozyme expression driven by the RNA polymerase III promoter was about ten fold higher compared with those driven by the RNA polymerase II promoter. Using the RNA polymerase III promoter to produce the ribozyme, Yu *et al.* reported a high-efficiency inhibition of the HIV-1 gene expression in cell culture . Unlike RNA polymerase II transcript, RNA polymerase III transcript can be designed to localize either within the cytoplasm or the nuclei. However, these promoters cannot be regulated in a cell-specific manner, thus limiting their use in certain situations.

The potential for use of the hammerhead ribozyme against several types of diseases such as cancer, HIV, and other viral infection diseases has been tested in human culture cells without significant negative effects on the normal cell growth. As already mentioned, the hammerhead ribozyme can be delivered as a synthetic RNA or expressed from an expression vector. Thus, the hammerhead ribozyme can be used as a tool for gene therapy against a number of diseases.

FUTURE PROSPECTS

Over the last few years, tremendous progress toward using short DNA molecules as therapeutic agents has been made. Additionally, a substantial transition from laboratory applications to clinical applications is being observed. Clinical trials with oligonucleotides have already been initiated, and currently more than a dozen commercial enterprises selectively focus their attention on the development of short DNA molecules for therapeutic use.

Results of studies on chromatin structure, DNA methylation, and alternative DNA structures are expected to make particular DNA targets more accessible to exogenous oligonucleotides or their analogs. Covalent linking and cooperative interaction between multiple ligands will help to create more stable complexes, and pharmacological studies in animal models will answer questions regarding the pharmacodynamics, pharmacokinetics, toxicity, cellular uptake, and stability of these oligonucleotides.

Thus, oligonucleotide technology is emerging as a powerful tool with a growing number of applications. With improvements in intracellular stability, delivery methods, and oligonucleotide design, and with rapid advances in our knowledge of gene sequences oligonucleotide technology is certain to become a major player in gene therapy.

4

New Food Crops

Throughout history, plant breeders have sought to genetically modify food crops to improve yield and increase resistance to disease and plant pests. Initially these improvements were achieved by selecting seed from superior plants and reproducing these with continual selection and breeding. Traditional breeding methods have increased corn and wheat yields by approximately 100% over the last half century. However, traditional plant breeding methods are slow and unpredictable. To introduce a desired gene or set of genes by conventional breeding methods requires a sexual cross between parental lines followed by repeated backcrossing between the hybrid off spring and one of the parents until progeny with the desired characteristics are obtained. Genes are only accessible from plants that can be sexually crossed, and many genes besides the desired gene (s) will be transferred.

Biotechnology provides an opportunity to overcome some of the limitations of traditional breeding by enabling plant geneticists to identify and clone specific genes encoding desirable traits, such as protection against insect pests, and to introduce these genes selectively into already useful varieties of plants. Sexual compatibility is no longer a limiting factor for transfer of desired traits. The transformation process is faster and more efficient because successfully transformed plants can readily be identified. Numerous traits are being assessed for their potential to yield products with the ability to (1) protect plants against various insect, fungal, and viral pests and plant pathogens; (2) provide selectivity to preferred herbicides; (3) improve agronomic performance such as crop yields; (4) increase nutritional value of food for humans and farm animals; (5) reduce naturally occurring toxicants, antinutrients, or allergens; (6) modify the ripening process to improve the flavour of fruits and vegetables; (7) use plants as factories to make environmentally friendly biodegradable polymers for packaging materials; and (8) use plants to produce pharmaceutical products more cost effectively, and so on.

Since the initial reports of the first genetically modified plants 15 years ago, nearly all agronomically important crops have been genetically modified. By the end of 1995, more than 70 different crop species had been transformed. At least 56 different crop have been planted in at least 34 countries and grown in more than 15,000 individual field sites. According to the U.S. Animal and Plant Health Inspection Service (APHIS), the most frequently field-tested traits are insect protection, virus protection, fungal resistance, herbicide tolerance, and food quality enhancements. The number of transformed crops and field trials will continue to expand as more new crop varieties are developed through biotechnology for eventual entry to the marketplace.

As shown in Table at least 34 genetically modified plants have successfully completed regulatory review by the appropriate regulatory agencies. Genetically modified plant were grown commercially in 1996 on approximately 7 million acres in various world areas with approximately 30 million acres planted in 1997.

Biotechnology provis plant breeders the opportunity to develop new varieties of food crops more efficiently and with greater potential benefit than has been possible with conventional breeding practices. As with any technological innovation, there must be assurance that the technology will deliver food "as safe as" that developed through traditional breeding programs. This chapter addresses the safety assessment strategies for new varieties of food products developed through biotechnology. The approaches are consistent with the guidance developed by various international organizations such as OECD (1993, 1996 1997), FAO/WHO (1996), and WHO (1991, 1995). An example of the application of these strategies to assess the safety of a genetically modified food plant.

To understand how safety assessment approaches have evolved, it is instructive to review briefly how new traits are introduced into food crops through conventional breeding compared with biotechnology.

TABLE 4.1 : EXAMPLES OF PLANT BIOTECHNOLOGY PRODUCTS THAT HAVE SUCCESSFULLY COMPLETED REGULATORY REVIEW IN AT LEAST ONE COUNTRY

Company	Genetic Trait
AgrEvo Canada, Inc.	Glufosinate-tolerant canola
	Glufosinate-tolerant corn
	Glufosinate-tolerant soybean
Agritope, Inc.	Modified fruit-ripening tomato
Asgrow Seed Co.	Virus-resistant squash I
	Virus-resistant squash II
Bejo–Baden	Male sterility/glufosinate-tolerant chicory
Calgene, Inc.	Flavr Savr™ tomato
	Bromoxynil-tolerant cotton
	Laurate canola
China	Virus-resistant tomato
Ciba Seeds	Insect-protected corn
Cornell U./U. of Hawaii	Virus-resistant papaya
DeKalb Genetics Corp.	Glufosinate-tolerant corn
	Insect-protected corn
DNA Plant Technology	Improved ripening tomato
DuOont	Sulfonylurea-tolerant cotton
	High-oleic-acid soybean
Florigene	Carnations with increased vase life

	Carnations with modified flower colour
Monsanto	Glyphosate-tolerant soybean
	Improved ripening tomato
	Insect-protected cotton
	Insect-protected Potato
	Glyphosate-tolerant cotton
	Glyphosate-tolerant canola
	Insect-protected corn
	Glyphosate-tolerant corn
Mycogne	Insect-protected corn
Northrup king	Insect-protected corn
Plant Genetic Systems	Male sterile oilseed rape
	Male sterility/glufosinated-tolerant corn
University of Saskatchewan	Sulfonylurea-tolerant flax
Zeneca/Petoseed	Improved ripening tomato

FOOD CROPS: WITH BIOTECHNOLOGY

Coventional Breeding

Genetic modification of plants has been practiced for hundreds of years with considerable success by plant breeders. Plant breeding has become a very sophisticated branch of applied genetics. Breeders have developed elegant procedures for crossing plants to introduce and maintain desirable traits such as increased yield and resistance to disease. With conventional breeding, cultivars highly adapted to cultivation can be improved by crossing them with closely related highly adapted cultivars to combine the desired features of the parents. If the parents are highly adapted to cultivation, their progeny tend to be highly adapted also. However, if a desirable trait is not available in highly adapted cultivars, the breeder will cross-adapt cultivars with cultivars from different geographic areas, more primitive varieties, or wild species. The greater the diversity in genetic material of the parents, the greater the chance that undesirable characteristics (genes) will be introduced into the progeny. The progeny may be an "offtype," which means it has less desirable agronomic properties, such as stunted growth or poor yield, than parent cultivars. Eliminating plants with undesirable traits while retaining plants with desired features may require many backcrosses carried out over several generations. The greater the difference in genetic content of the parents, the more difficult sexual crossing can be. *In vitro* procedures such as embryo culture and protoplast fusion have made it possible to cross parents that may not be sexually compatible.

Biotechnology

Biotechnological methods do not replace conventional breeding practices but can facilitate the introduction of desirable traits into the plant genome more efficiently and with greater precision. Unlike traditional breeding in which thousands of genes may be introduced into progeny from their parents, only one or a few genes are typically introduced using biotechnologial techniques. The source of the desired trait the breeder wishes to introduce into new crop varieties is no longer limited to those from sexually compatible species. This greatly expands the opportunity to introduce new traits to improve crop varieties.

The techniques used to introduce new traits into food crops vary depending on the kind of plant being transformed. Seed plants have been divided into two subclasses: monocotyledonous plants (monocots), whose seeds have a single cotyledon (meaning "seed leaf"), and dicotyledonous plants (dicots) or those with two cotyledons. Dicots (broadleaf plants like tomato, cotton, and soybean) were the first seed plants to have genes introduced via biotechnology. The first genetically modified plants were produced in 1982 using *Agrobacterium tumefaciens*, a bacterium that can transfer a portion of its own DNA (T-DNA) into the genome of plants. The disease-causing sequences from the *Agrobacterium* T-DNA have been deleted followed by insertion of DNA that encodes for a desired trait such as insect protection. Regulatory singals are added to enable the gene to function optimally in the plant. Border sequences in the plasmid delineate the desired genes that will typically be inserted in the plant genome. The T-DNA is incorporated into a plasmid derived from *Escherichia coli*, which is transferred to *A. tumefaciens* via a conjugation process. *A. tumefaciens* containing the engineered plasmid is incubated with selected tissue from the host plant, and the desired genes are stably inserted into the plant chromosome. The inserted genes can then be transferred to new plant varieties using traditional breeding methods.

Monocots, which are represented by agronomically important cereal crops such as wheat, rich, and corn, were not initially amenable to transformation with *A. tumefaciens*. Recently, more aggressive strains of *A. tumefaciens* with broader host ranges have been developed that have been used to introduce genes into some cereal crops. In addition, techniques such as protoplast transformation and particle bombardment have been used to transfer DNA directly into cells where it is stably inserted into the plant genome. These techniques have been particularly valuable in monocot plants for which the *Agrobacterium* transformation method was not effective or efficient. In particle bombardment, very small metal beads (1-μ diameter) are coated with DNA and shot from a "gun" into the target monocot cell. Some of the plant cells will incorporate the desired gene (s) into their genome.

Whether *Agrobacterium* transformation or particle bombardment is used to introduce genes into plant cells, only a small percentage of eligible plant cells will be successfully transformed. To identify the transformed cells in culture, genes for selectable markers have been included with the genetic information inserted into plant cells. The marker genes provide resistance to antibiotics, herbicides, and other substances added to the cell culture to inhibit the growth of nontransformed cells. Plant cells that survive have been successfully transformed and can therefore be identified and regenerated into whole plants.

As with conventional breeding, offtypes are typically discarded. Progenies with normal agronomic properties that express the desired phenotype or trait are backcrossed with commercial crop varieties to generate seed bearing the new trait. The chances of success in identifying a progeny with the desired trait are much higher with recombinant DNA techniques because the breeder knows that the trait was successfully inserted into the plant genome. With conventional breeding, many generations of backcrossing may be required to identify a progeny with the desired trait.

CONVENTIONAL BREEDING

During this century, our food supply has steadily improved in quality, variety, nutritional value, safety, and economy through the use of conventional breeding techniques to improve food crops. The history of safe use of new varieties of food crops developed by classical breeding techniques is based on several factors: (1) confidence and experience with the procedures used to generate new crop varieties; (2) knowledge of the composition of the food crop, including important nutrients and toxicants, if present; and (3) observation of the agronomic properties of new crop varieties to eliminate those with undesirable properties. Of the thousands of new fcrop vaieties that have been developed during this century via traditional breeding, only a very limited number of new varieties have presented safety concerns. The more well-known examples are (1) increased psoralens in certain varieties of celery that caused photodermatitis in food handlers; (2) increased glycoalkaloid content in the Lenape variety of potatoes (glycoalkaloids can cause gastrointestinal discomfort); and (3) increased cucurbatin levels in vegetable squashes that leave a bitter taste (IFBC 1990). In these three examples, the food crops contained endogenous toxicants affording protection against plant pests. The levels of these endogenous toxicants were inadvertently increased in these new varieties. These exampled have caused plant breeders to monitor new varieties of food crops that naturally contain potentially harmfull toxicants or antinutrients more carefully to be certain that levels of these substances are within acceptable limits. For example, the United States and Canada have set acceptable limits for glycoalkaloid levels in potatoes that new potato varieties must meet.

Regulation of New Crop Varieties

Conventional breeding

In the United States, there are no premarket regulatory requirements governing the introduction of new crop varieties developed through conventional breeding (with the exception that the variety must not exceed standards set for the level of certain natural toxicants such as glycoalkaloids in potatoes). Food safety is assessed through postmarketing measures under the Food, Drug, and Cosmetic Act. This practice is based on the long history of safe introduction of new crop varieties developed through conventional breeding. As discussed earlier, plant breeders monitor the quality of new plant varieties before they are introduced into commerce. In Europe and for some crops in Canada, new varieties of food crops must be registered with the government. This is not done for safety reasons but more as an assurance to the farmer that the new variety will perform at least as well as commercial varieties in the field. In the United States, the market place determines the performance acceptability of new crop varieties.

Biotechnology

Because biotechnology is relatively new, there is considerably more regulatory over sight for the introduction of new frop varieties developed through its approach. In the United States, the regulatory authority to ensure the safety of food and feed products derived from plant biotechnology resides within the Food and Drug Administration (FDA). The U.S. Department of Agriculture (USDA) has the authority to ensure that genetically modified plants will not become plant pests. The Environmental Protection Agency (EPA) has the authority to evaluate the safety of plants that have been genetically modified for protection against plant pests such as insects, fungi, bacteria, and viruses. The EPA also regulates herbicides by establishing herbicide tolerances for plants genetically modified to be herbicide tolerant.

The movement and release of genetically modified plants is regulated by USDA under the Federal Plant Pest Act and the Plant Quarantine Act. Permits or notifications must be filed with USDA to obtain approval for field testing of new varieties of genetically modified plants under development. A determination that the genetically modified plant is not a plant pest (e.g., that the plants do not pose a risk to the environment or production agriculture) must be obtained before market introduction.

The FDA has the regulatory authority to ensure the safety and wholesomeness of food and feed products, including those derived from genetically modified plants. The FDA has adopted a decision-tree approach to ensure safety of products derived from new varieties of food and feed developed by both traditional and newer genetic modification methods, including biotechnology. Under the Food Drug and Cosmetic Act [21 CFR 402(a) 1] the FDA has the authority to take regulatory action against a new variety of food if the genetic modifications would render the food "ordinarily injurious" to human health (IFBC 1990). Therefore, the FDA uses the same postmarket food adulteration approach and regulations for these products as are used for food and feed products derived from traditionally bred plant varieties. It is recommended that developers of genetically modified food consult with the FDA before commercialization, and this has been done with all genetically modified food products that are in the marketplace. The FDA has completed consultations on at least 29 different genetically modified crop plants to date.

The Federal Insecticide, Fungicide and Rodenticide Act (FIFRA) provides the EPA with the authority to regulate plants with bioengineered pesticidal traits. The EPA has approved at least seven different genetically modified plants with introduced pesticidal traits. These products are also reviewed by the FDA and USDA.

In regard to regulation of genetically modified plants in other world areas, Health Canada regulates food safety, and Agri-Food Canada regulates feed and environmental safety as well as registration of specific plant varieties for certain crops. In Japan, the Ministry of Health and Welfare regulates food safety, whereae the Ministry of Agriculture, Food, and Fisheries regulates feed and environmental safety. In the European Union, environmental assessments are conducted before placing a product on the market, as outlined in the 90/220 EEC regulations. The recently authorized Novel Foods Regulation governs food safety in the European Union and replaces individual country food regulations that were in place in the United Kingdom, Denmark, and the Netherlands. A Novel Feed Regulation is under consideration for overseeing feed safety under the current 90/220 EEC process. Many other countries either have, or are in the process of developing, regulations for plant biotechnology products. The number of genetically modified plants that have successfully completed regulatory review in various countries include 23 in Canada, 5 in the European Union, 15 in Japan, 3 in Mexico, 2 in Argentina, 1 each in Australia and Brazil.

New Crop Varieties through Biotechnology

The safety issues that have been raised for genetically engineered plant products are similar to those for new varieties of plants derived from conventional breeding. For example, progenies derived from conventional breeding as well biotechnology may have progenies with altered agronomic properties when compared with the parents. Varieties with altered agronomic properties are usually readily identified in field trials. It has been suggested that biotechnology may inadvertently cause the production of new toxicants through insertional mutagenesis events that activate dormant biosynthetic pathways. This scenario seems very unlikely—particularly for those crops where there has been considerable experience with traditional breeding. It is more likely that an insertional event would result in the increase or decrease in the expression of already recognized toxicants, which has very infrequently been observed in conventional breeding, as discussed above.

In conventional breeding, the genes bearing the desired trait have often not been identified. The same applies to the protein expression products of these introduced genes. Introduction of genes through biotechnology procedures is much more precise, for the genes have been defined before their introduction. However, unlike conventional breeding, introduced genes can be obtained from almost any source, not just sexually compatible relatives of the food crop. Therefore, as part of the safety evaluation, regulatory agencies have required a molecular characterization of any gene introduced into food crops. In addition, the protein expresssion product of the cloned gene must be characterized as to its function, specificity of action, and safety. There are no particular concerns about the safety of the genetic material itself. Genetic material from living organisms is made from the same four nucleotide building blocks; the only difference between genes is the nucleotide composition. The human gut at any one time has been estimated to contain hundreds of milligrams of DNA from ingested food and mucosal cells sloughed into the gastrointestinal tract. These DNA molecules are efficiently degraded by nucleotidases during digestion. The contribution of genetic material from genes cloned into food is trivial compared with the other sources of DNA in the gut. It was concluded that the introduced genes (DNA) in food products pose no more health risk to consumers than the rest of the DNA ingested from food sources.

Molecular Characterization

For new plant varieties developed through biotechnology, the source of the gene introduced into the plant must be identified. The transformation system used to insert the gene into the plant genome must be defined as well as the number of copies of inserted genes and the integrity and stability of the genetic insert. For genes coding for pesticidal proteins, the EPA requires the following information: (1) description of the vectors, (2) identity of organisms used for the cloning of the vectors, (3) description of the methodologies used to clone the vectors. (4) vector description (size in kilobases), (5) restriction endonuclease sites, (6) location and function of all relevant gene segments, (7) the final delivery system, (8) description of gene segments transferred to the plant, (9) description of whether the inserted genes are expreseed constitutively or inducibly, (10) localization and expression of the pesticidal substance in plant parts, and (11) estimation of the gene copies inserted, and so on.

Substantial Equivalence

A general consensus has existed among regulatory agencies in major world areas regarding the use of "substantial equivalence" as an approach to assess the safety and acceptability of genetically modified crops. This concept has been adapted in part from procedures that plant breeders have followed to monitor the acceptability of new varieties of food crops developed through conventional breeding.

According to OECD, "the concept of substantial equivalence embodies the idea that existing organisms used as food or food source can serve as a basis for comparison when assessing the safety of human consumption of a food or food component that has been modified or is new. If a new food or food component, is found to be substantially equivalent to an existing food or food component it can be treated in the same manner with respect to safety, keeping in mind that the establishment of substantial equivalence is not a safety or nutritional assessment in itself, but an approach to compare a potential new food with its conventional counterpart".

The use of substantial equivalence is considered as practical approach to assess the safety of new varieties of food. Safety is defined as a reasonable certainty that no harm will result from intended uses under the anticipated conditions of consumption.

The following are the three possible outcomes relative to assessing the substantial equivalence of a genetically modified food or food component to a conventional counterpart.

1. The genetically modified food or food ingredient is substantially equivalent to the conventional counterpart;
2. The genetically modified food or food ingredient is substantially equivalent to the conventional counterpart with the exception of certain well-defined differences;

   or
3. The genetically modified food or food ingredient is not substantially equivalent to the conventional counterpart.

Examples of new varieties of food crops that could be considered substantially equivalent to conventional counterparts are virus-resistant plants produced be introduction of viral coat protein (viral protein is aready present in plant tissue of conventional infected plants). Alternatively, if the food is processed removing any added traits and the composition has not changed (e.g., processed canola oil), the oil would be considered substantially equivalent to its conventional counterpart.

The second outcome listed above will apply most genetically modified food crops. For example, many genetically modified crops will be substantially equivalent to conventional counterparts with the exception of a new trait that imparts a desired characteristic such as pest resistance. When this is the case, further safety assessment should focus on the new trait or well-defined differences. For some genetically modified crops or food components derived therefrom, it may not be possible to demonstrate substantial equivalence to a conventional counterpart, either because differences are not sufficiently well difined or because there is no appropriate counterpart with which to make a comparison. The absence of substantial equivalence does not imply that the genetically modified crop is any less safe. The safety assessment should focus on the nature of the changes.

Assessment of Substantila Equivalence

An assessment of substrantial equivalence should focus on a comparison of agronomic characteristics combined with compositional analysis that includes key nutrients as well as toxicants and antinutrients that have potential health singificance.

1. Agronomic traits are a good starting point for evaluating substantial equivalence of genetically modified plants to their coventional counterpart. These traits will normally be examined as an integral part of the development program leading to a new food plant variety. Agronomic traits are those characteristics measured by plant breeders for a given crop. For example, in the case of potatoes these may include yield, tuber size and distribution, dry matter content, and disease resistance. Agronomic properties may vary depending on local environmental conditions. Therefore, the genetically modified variety should be grown in the same geographical regions in which it will be grown commercially.

2. Key nutrients are those components in a particular food product that may have a substantial health impact in the overall diet. These may be major constituents(fats, proteins, carbohydrates) or minor components (essential mineral, vitamins). Critical nutrients to be assessed may be determined, in part, by knowledge of the function and expression product of the inserted gene (e.g., if an inserted gene expresses an enzyme that is involved in amino acid biosynthesis, the amino acid profile should be determined). Introduction of an invertase into potatoes could influence the carbohydrate metabolism; therefore, the amino acid profile should be measured. Examples of the analysis of key nutrients in different genetically modified plants have been published.

   Given difference among consumption patterns and practices in various cultures and societies, the key nutrients to be examined may differ in various countries. The critical nutrients to be addressed should be determined using consumption data for the target region. For example, in Denmark, potatoes provide an important source of vitamin C in the diet (35%), not because of their high vitamin C content (20 mg/100 gm). In the United States, potatoes are not as important a source of vitamin C owing to lower potato consumption and the availability of other dietary sources of this vitamin.

3. Critical toxicants antinutrients are those compounds known to be inherently present in a crop variety whose potency could have an impact on health if their levels were increased significantly (e.g., solanine glycoalkaloids in potatoes, trypsin inhibitors in soybeans). Knowledge of the biologic function of the protein expression product of the inserted gene could influence the decision about which toxicants or antinutrients to examine.

The analysis of important nutrients and toxicants or antinutrients should use validated or standard methods where available (e.g., methods from the Association of official Analytical Chemists [AOAC] or other recognized bodies).

When the genetically modified line is compared with the parental line or conventional varieties, the varieties should be grown under similar environmental and agronomic conditions. If there are no statistically significant differences in measured parameters between the genetically modified line and the parent or conventional lines, then the genetically modified variety can be considered substantially equivalent to the parent or conventional variety. If statistically significant difference are observed in measured parameters, then a comparison can be made with values available in literature or other sources. If the parameter for the genetically modified line is within the normal range for conventional varieties, further evaluation is not warranted. If the parameter for the genetically modified line is outside the normal range, further evaluation may be needed. A toxicological or nutritional evaluation may be needed to assess whether the difference between conventional and genetically modified lines for a given parameter is biologically meaningful and requires further investigation.

In the future, genetically modified lines for sexually compatible crops will be crossed using traditional breeding techniques. If substantial equivalence has been demonstrated for (1) corn with a gene producing insect protection and (2) will produce a new variety that is likely to be substantially equivalent to the parents. If there are no expected interactions between the two traits, then no additional evaluation will be needed. Potential genetic interactions will need to be considered on a case-by-case basis. For example, if two independent modifications to the same metabolic pathway are combined by traditional breeding, further analysis of the products of the metabolic pathway may be warranted.

Gene Expression Product

Many genetically modified food crops have been shown to be substantially equivalent to conventional crops with the exception of the introduced trait (s) that may impart one or more characteristics such as pest resistance, selectivity to preferred herbicides, modification of the ripening process, and so forth. For these examples, the safety assessment should focus on the introduced trait, the protein expression product of the cloned gene. The developer should do the following:

1. Define the biological function, specificity, and mode of action of the protein. If the protein is an enzyme, assess the potential effects of the enzyme on metabolic pathways and levels of endogenous metabolites based on its mode of action and specificity.
2. Compare the amino acid sequence of the protein to known sequences in protein databases (GenPept, SwissProt, PIR) to determine if the protein has sequence homology to food proteins, toxins, or allergens.
3. Assess the inherent digestibility of the protein in vitro with simulated gastric and intestinal protease preparations.
4. Determine the level of expression of the protein in the food. This effort should focus on the raw agricultural product or a specific processed food component (e.g., oil), as appropriate.

Proteins in Foods

The following criteria are important in assessing the safety of a protein introduced into food or food components:

1. The protein has a history of safe consumption in other food crops.
2. The protein is functionally and structurally related to proteins with a history of safe consumption in food.
3. The biological function and specificity or mode of action of the protein raises no safety concerns.
4. The amino acid sequence of the protein is not similar to known protein allergens.
5. The protein is not derived from a food source with a history of allergy.
6. The amino acid sequence of the protein is not similar to known protein toxins or antinutrients.
7. The protein is susceptible to degradation by digestive enzymes.

For certain insect control protein such as the *Bacillus thuringiensis (B. t.)* family of proteins, regulatory agencies such as the EPA require administration of the protein as a single oral high dose to mice. The scientific rationale for an acute test in mice is that known protein toxins generally manifest their toxicity via acute mechanisms. The *Bt.* proteins have a long history of safe use (EPA 1988); as new forms of *Bt.* proteins are introduced, the EPA has stated that an acute dosing test in mice will provide assurance that the new *Bt.* protein varieties are safe. Such studies have been conducted for the Cry3A protein expressed in potatoes, the Cry1Ab protein expressed in corn, and the CryIAc protein expressed in cotton and the tomato. Enzymes that provide selectivity against herbicides or function as selective markers have also been tested in mice. Acute testing may be recommended if the protein is derived from plants, or microorganisms that have no history of consumption. Acute toxicity testing of proteins is not normally needed if the protein (1) is not present in the food (e.g., removed during processing, as in vegetable oils), (2) has a history of consumption in food, and (3) is closely related functionally and structurally to proteins with a history of consumption in food. If safety testing of proteins is undertaken, it may be possible to isolate and purify the protein from the plant. However, this may not be practical if the protein is expressed in small amounts in plant tissue. Alternatively, the protein could be produced in an appropriate fermentation system generating a protein that is biochemically equivalent to the protein produced in the plant.

Proteins that meet the safety criteria listed and nontoxic if administered to mice are considered "as-safe-as" other proteins naturally present in foods. No further safety or nutritional assessment of the new variety of food crop is necessary.

Further Safety Evaluation

Some introduced proteins may require further safetyand nutritional evaluation if they meet any of the following criteria:

1. The protein is functionally and structurally related to proteins that are known toxins or antinutrients.
2. The protein is derived from a food with a history of al-

lergy.

3. The protein is structurally similar to known protein allergens.
4. The protein is not degraded by digestive enzymes.
5. The biological function of the protein has not been characterized and there are no structurally related proteins identified in protein databases.

If the introduced proteins meet one or more of the criteria listed above, additional testing may be warranted on the basis of a case-by-case assessment.

Testing Recommendations

Certain antinutritional proteins such as lectins or protease inhibitors are naturally present in plants and provide protection against insect pests. There has been an interest in inserting these proteins in food crops to provide and alternative to insecticides for control of insect pests, although the safety implications of such transformations have raised concerns. Some plant lectins exert antinutrient effects when fed to animals by binding to the brush-border epithelium of gut cells and thus disrupting nutrient absorption. If a lectin were to be introduced into a food crop to enhance protection aginst insect pests, it should be fed to animals to assess whether it acts as an antinutrient. Assessing its potential for binding to brush-border epithelium would be part of the safety evaluation. If the introduceed lectin had antinutrient properties, a "no-effect level" for antinutrient effects would have to be determined in animal feeding studies and compared with potential human exposures to determine if an adequate safety margin existed. Human nutritional studies may also be appropriate. Because some lectins are known food allergens, assessment of potential allergenicity will also be necessary, A similar testing scheme for protease inhibitors can be envisioned. Any general safety questions regarding the introduced protein could be addressed by animal feeding studies designed to assess toxicity and nutritional endpoints. For proteins containing common amino acids, no additional testing for mutagenic, carcinogenic, or teratogenic potential is indicated because there is no evidence that proteins are genotoxic or that feeding proteins such as food enzymes have ever directly produced carcinogenic or teratogenic effects in laboratory animals.

Testing for Allergenicity

Further testing of an introduced protein for potential allergy would be indicated if (1) the protein were derived from a food source with a history of allergy or (2) the amino acid sequence of the protein matched (at least eight contiguous identical amino acids) that of a known protein allergen. The immunogenic potential of the introduced protein should be tested in one of the various solid-phase immunoassays such as the radio allergosorbent test (RAST) or RAST inhibition assay or the enzyme-linked immunosorbent assay (ELISA). Solid-phase immunoassays use IgE fractions of sera from individuals confirmed allergic to the food from which the gene coding for the introduced protein was derived. Where *in vitro* tests are negative or equivocal, then *in vivo* skin prick tests could be carried out. If no positive response was detected in the prick test, then double blind placebo controlled food challenges with patients known to be allergic to the food could be carried out under controlled clinical conditions. The in vivo studies would only be warranted if the gene were derived from one of the following eight food groups that account for more than 90% of all food allergies: eggs, milk, fish, crustacea, peanuts, soybeans, wheat, and tree nuts.

The testing scheme to detect allergens summarized above works effectively as demonstrated in the case of the Brazil nut 2S storage protein. This protein was introduced into soybeans to increase the sulfure amino acid content and there by improve the bean's nutritional value for use in animal feeds. Because there are a small number of individuals to see if their sera contained IgE that would crossreact with the Brazil nut 2S storage protein. Sera from eight out of nine Brazil-nut-allergic individuals reacted with this storage protein. Development of this soybean line containing the introduced Brazil nut storage protein was terminated. If the developer has wished to proceed with commercialization of this genetically modified soybean, all foods derived from this soybean would have had to be labeled as containing Brazil nut protein.

For many genetically modified plants, the protein expression product will be derived from a gene source that has no history of inducing allergy. If the amino acid sequence of the introduced protein does not show sequence homology to known allergens (see above), the physicochemical properties of the protein should still be assessed to determine if it shares the profile for allergens. These properties include (1) stability to food processing conditions (high temperature and pH changes that denature food proteins), and (2) resistance to gastric acidity and digestive proteases. The level of expression of the introduced protein in food should also be determined because allergenic proteins often comprise a significant proportion of the total protein in that food.

Regarding digestibility of proteins, it is well established that protein allergens tend to be resistant to digestion and generally consititute a significant percentage of the total protein (1–18%) in the food. Allergens are often stable to heat processing. These processing and the hostile environment of the gut to elicit immunologic reactions in the intestinal mucosa. However, there are undoubtedly examples of proteins in food that represent a small percentage of total protein, are not digestible, and do not elicit food allergy. According to allergenicity experts, there is currently no validated *in vivo* or *in vitro* model to predict allergenicity potential of proteins. There is therefore a need to develop a predictive model for allergenicity.

Marker Genes

As stated earlier, marker genes are added to the genetic meterial inserted into plants to help identify successfuly transformed plant cells. The use of marker genes and their protein expression products in plant biotechnology has raised the following safety concerns: (1) potential horizontal transfer of the marker gene from plant cells to gut microflora that might reduce the efficacy of therapeutic antibiotics, (2) potential toxicity of the protein expression products of the marker gene, or (3) compromised antibiotic efficacy due to expression of the antibiotic marker gene product in the food. After detailed scientific review, it was concluded that the potential for horizontal gene transfer from plants to gut microflora is vanishingly small. Introduced genes are stably incorporated into the genome of plants, and there are no known mechanisms for direct transfer of genes from plants to microorganisms. Moreover, the DNA released from the plant during digestion is rapidly degraded. This degradation occurs well before the plant material reaches the lower intestine, cecum, and colon, where gut microflora are found. The potential for the protein expression product of the antibiotic marker gene to compromise antibiotic efficacy is limited by (1) the digestibility of the expressed protein, (2) low expression of the protein in food, or (3) lack of available cofactors such as adenosine triphosphate (ATP) in the gastrointestinal tract that are required for antibiotic inactivation. The safety assessment of the protein expression product of marker genes should focus on the approach outlined earlier for protein expression products of introduced genes (new traits).

Genetically Modified Plants

To date, there have been few, if any, examples of genetically modified plants that are not considered substantially equivalent to conventional counterparts (with the exception of well-defined differences in some cases). As discussed earlier, with future developments in biotechnology, products may be developed that have no conventional counterpart for which substantial equivalence can be assessed. These products could arise by transfer of genomic regions that have only partially been characterized. Genomics is the study of all the genes of an organism and their organization into chromosomes, and it includes researching the linked between gene structure and functions, or expression, that are the foundations for bioengineering new plant products. Desirable agronomic traits may result from the linkage of several genes functioning together to produce effects such as resistance to drought or alkaline soil conditions. Such a development would allow plants to survive and grow under harsh environmental conditions, and millions of acres of land that cannot be currently used to grow food crops would be available for cultivation of genetically modified food crops.

In the future, there may be novel foods derived via biotechnology that have intended benefits, but further safety and nutritional evaluation will be needed before these foods can be marketed. In othere cases, unintended changes in a new food crop may occure that could require further evaluation to understand what was changed and its significance to the safety and nutritional value of the food crop.

Toxicological and nutritional studies may be indicated in some cases to resolve safety and nutritional questions of food crops that are not substantially equivalent to their conventional counterparts. If animal feeding studies are considered neccessary, their objective must be clear and the experimental design carefully planned to avoid nutritional problems that may confound date interpretation. There may be a need to modify existing protocols for testing whole food or food components— in particular by providing adequate nutrition with respect to diet. Nutritional imbalance may mask toxic effects. Incorporation of high levels of a food into the diet may also result in nutritional deficiencies that could cause adverse effects unrelated to the food being tested. The usual concept of safety margins may not be applicable because it is often not possible to feed a whole food at a high enough level to obtain anything close to a hundredfold margin of safety. The testing of specific components or extracts of a novel food may present an option for addressing safety and nutritional testing of the whole food. For certain foods, animals are not good models for predicting safety to man because animals exhibit species-specific sensitivity to the food. For example, dogs experience transient paralysis from consumption of macadamia nuts or develop fatal cardiac arrhythmias owing to sensitivity to theobromine in chocolate; male rats of one particular strain develop cardiomyopathy when fed diets high in vegetable oils. Feeding animals high levels of various foods in the diet has produced anemia (rats and dogs fed onions), enlarged cecums (rats fed potatoes), mucosal lesions in the stomach (rats fed tomatoes or chilli powder), pulmonary emphysema (rats fed beans) and reducal breeding performance and survival (rats fed wheat flour), If nutritional quality is the important issue, it may be appropriate to go directly to nutritional studies in human volunteers rather than to use animal models that may have limited relevance to man. The industry currently uses "sip and spit" tests to evaluate the organoleptic properties of new varieties of vegetable crops as part of the quality assessment. Once the initial safety assessment has been completed for a nonsubstantially equivalent food product, sensory or nutritional evaluations, or both, may be recommended to ensure the quality and acceptability of the product for consumers.

CONCLUSION

Biotechnology provides plant breeders the opportunity to develop new varieties of food crops more efficiently and with greater potential benefit than has been possible with conventional breeding practices. To provide assurance that this technology will generate food "as safe as" that produced by traditional breeding programs, safety assessment strategies have been developed for products of plant biotechnology that have generally been accepted by international regulatory groups (FAO/WHO 1996). This strategy includes comparison of the agronomic properties and important nutrient toxicant composition of genetically modified crops is said to be "substantially equivalent" to its conventional counterpart, and the safety and nutritional assessment is completed. Many genetically modified foods will be substantially equivalent to conventional varieties with the exception of one or more introduced traits. The safety assessemt will then focus on the introduced trait to determine if the protein expression product meets the criteria for being "as -safe-as" proteins already in the food supply.

If protein expression products are derived from plants with a histroy of allergenicity; have a structural or functional similarity to known allergens, toxins, or antinutrients; or have functions that may alter the nutritional quality of the plant, then additional safety and nutritional studies may be needed. In the absence of validated assays to predict potential allergenicity, comparison of amino acid sequence homology of the protein with known allergens and assessment of its physicochemical properties are the best tools currently available to predict potential allergenicity.

As biotechnology develops, future products may be developed that have no conventional counterpart for which substantial equivalence can be assessed. These varieties could be designed to have improved agronomic properties or provide important health and nutritional benefits. The safety and nutrition assessment of these crops should be carried out on a case-by-case basis using the strategies outlined above.

During the last few years, over 34 different genetically modified plant products have successfully completed regulatory review in countries around the world. The experiences gained in bringing these products to market have enabled developers and regulatory agencies to define the key safety questions and the science to answer the questions. This progress will help provide guidance to the development of improved and safe plant biotechnology products for the future.

5

Soil Biotechnology

Deterioration of soils and aquifers has become evident in the last few years as a result of inappropriate final disposal procedures for all sorts of waste materials. Oil exploitation, uncontrolled fuel spills, poor practices for final disposal of industrial wastes, overuse of pesticides, and operation of sanitary landfills are some causes of pollution of soils and groundwaters.

In the course of the last two decades a wide variety of technologies has been developed for clean-up operations of contaminated soils and aquifers. They can be classified in terms of their principle of operation: physicochemical, thermal and biological. Among the biological technologies bioremediation has evolved as the most promising one because of its economical, safety and environmental features since organic contaminants become actually transformed, and some of them are fully minerlized.

The success of bioremediation techniques is directly related to the metabolic capability of involved microorganisms and can be affected by the surrounding microenvironment. This chapter describes the mechanism of bioremediation; an overview is presented of its virtues and weaknesses.

Soil and subsoil constitute a non-renewable natural resource that plays different roles, as described by Aguilar (1995):

1. Filtering medium during aquifer recharge
2. Protective layer of aquifers
3. Scenario of biogeochemical, hydrologic and food chain

processes

4. Natural habitat for biodiversity
5. Space for agricultural and cattle-breeding activities
6. Space for green areas to serve as sources for oxygen regeneration
7. Physical foundation for building construction
8. Sanctuary of the cultural reserve

The first four functions are the most important for the subject of this chapter, and treated as a whole they refer to what is commonly known as 'buffer capacity' of the soil. This buffering phenomenon constantly happens at recharge zones where rainfall migrates vertically downwards towards the aquifers. During seepage a large proportion of solid materials carried by the water are retained at the shallow layers and only water-borne dissolved chemicals seep downwards. Migration time depends on particle size distribution of soil; movemenet is faster through a fractured medium than through a granular geological material.

Seepage through the subsoil is delayed for compounds that do not migrate at the same rate as water. This frequently occurs in soils with a high organic matter content, such as clays in which organic compounds tend to be retained. There is also another important aspect related to particle size of soils. Clays are characterized by particles of small size (<2 μm); migration is therefore slower and the contact period among exogenic organic compounds and organic matter in the soil is longer, thus favouring the development of the sorption phenomenon. There the wide diversity of heterotrophic microorganisms involved in matter recycling start exerting their metabolic activity by using the existing organic compounds as carbon sources. The longer the substrate/microorganism contact time, the higher the possibility for degradation of the organic matter. The sorption and degradation phenomena of organics in the shallower geological material make it possible for water that continues its migration towards the aquifers to become free from exogenic compounds.

As a result of industrial activities spills commonly occur and water-insoluble organic contaminants seep into the soil. These to compounds, named NAPLs (non-aqueous phase liquids), have been classified in two types: those lighter than water (known as LNAPLs) and those denser than water (DNAPLs). Typical LNAPSs are petroleum hydrocarbons. their combustible products (such as gasoline. diesel and jet-fuel). benzene, toluence, ethylbenzene and xylenes (BTEX) used as industrial solvents. Other chlorinated industrial solvents, such as tertrachlorethylene, trichlorethylene, chloroform, carbon tetrachloride and methylene chloride, are examples of DNAPLs.

It is important to take into account the classification of contaminants in terms of their density because when LNAPLs reach the water table they tend to float on the water and to spread radially, whereas the DNAPLs continue their downwards path until bedrock is reached to arrest their movement. In terms of contamination effects, DNAPLs are more hazardous because they are capable of polluting the whole aquifer.

MICROORGANISM SURVIVAL IN ADVERSE CONDITIONS

Microorganisms possess wide biochemical versatility, which enables them to readily adapt to differnt microenvironmental conditions of pH, temperature and pressure, even extreme variations. The presence of high pollutant concentrations can be also regarded as extreme conditions. In these cases, contaminants induce a toxic effect on microbial activity to such a degree that the vital functions are inhibited. Microorganisms are capable of developing a certain tolerance to these adverse conditions and to become energy yielding for survival purposes. However, this happens only when microorganims have the genetic information available, or if they are capable of developing it, to allow the synthesis of enzymes that participate in the transformation of contaminants.

The organic-type contaminants can be used as carbon sources, thus achieving a reduction in their concentration and quite probably complete mineralization, i.e. degradation reaches the generation of carbon dioxide. This is highly desirable, although the presence of a cosubstrate might be required to support the energy-yielding activity while biotransformation of contaminants is achieved in different, and hopefully less toxic, chemical entities.

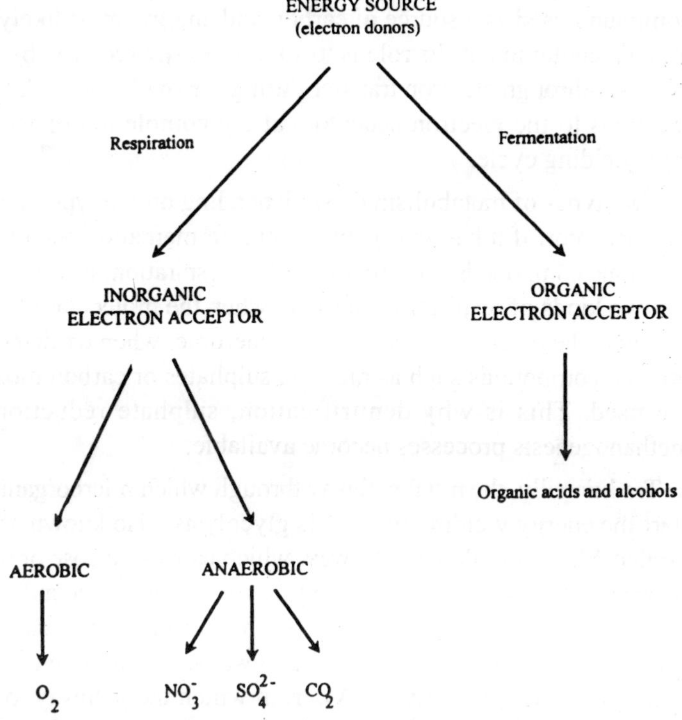

Figure 5.1 : *Route of metabolism depending on electron acceptor.*

Inorganic contaminants can only be transformed into different molecular entities; some are retained by the cells without becoming degraded. Reduction in the concentration of inorganic contaminants can be observed only when the microbial activity occurs in water where compounds move from the aqueous phase to the inside of the cells.

When reference is made to biodegradation, rather than simply referring to the microorganicsm, it is advisable to consider the enzymes which act as catalysts of the transformation reactions induced by the energy-yielding process. For this mechanism to occur the presence of an electron donor and of an electron acceptor is required as well as microenvironmental conditions suitable for synthesis and for the expression of catabolic enzymes. In the case of heterotrophic microorganisms, the electron donor will be the compound used as a source of carbon and energy, most likely the organic contaminant. Its role is to supply energy required by metabolism through electron transfer during the oxidation–reduction reactions to the electron acceptors at the completion of the energy yielding cycle.

Two types of metabolism exist, depending on the type of electron acceptor: if it has an organic origin, fermentation occurs; for inorganic compounds, the process will be respiration. In turn, there are two kinds of respiration: aerobic, when the molecular oxygen becomes the electron acceptor; and anaerobic, when oxidized inorganic compounds such as nitrates, sulphates or carbon dioxide are used. This is why denitrification, sulphate reduction or methanogenesis processes become available.

Traditionally, the nitial pathway through which microorganisms start the energy yielding process is glycolysis, also known as the Emden-Meyerhof–Parnas pathway, which carries glucose or other sugars to an intermediary such as pyruvate. Other compounds with different chemical characteristics (such as amino acids and fatty acids) have different degradation pathways, but all arrive at the same intermediary (acetyl CoA), from which the pathway to follow can be defined. When the respiration pathway is aerobic, three more degradative pathways are then followed: the citric acid cycle, electron transport chain, and oxidative phosphorylation. The presence of molecular oxygen is imperative for the last, which constitutes the most important mechanism for energy supply to the cellular activity; carbon dioxide is generated from the reaction. If this happens, it is assumed that the organic contaminants have become fully mineralized.

For degradation of organic compounds, the presence of enzymes is necessary; however, these are synthesized only when the cells have the specific genetic information and the substrate with which they interact are present in the required concentrations. As an example, mention can be made of the degradation of monoaromatic compounds indicative of contamination, such as gasoline,

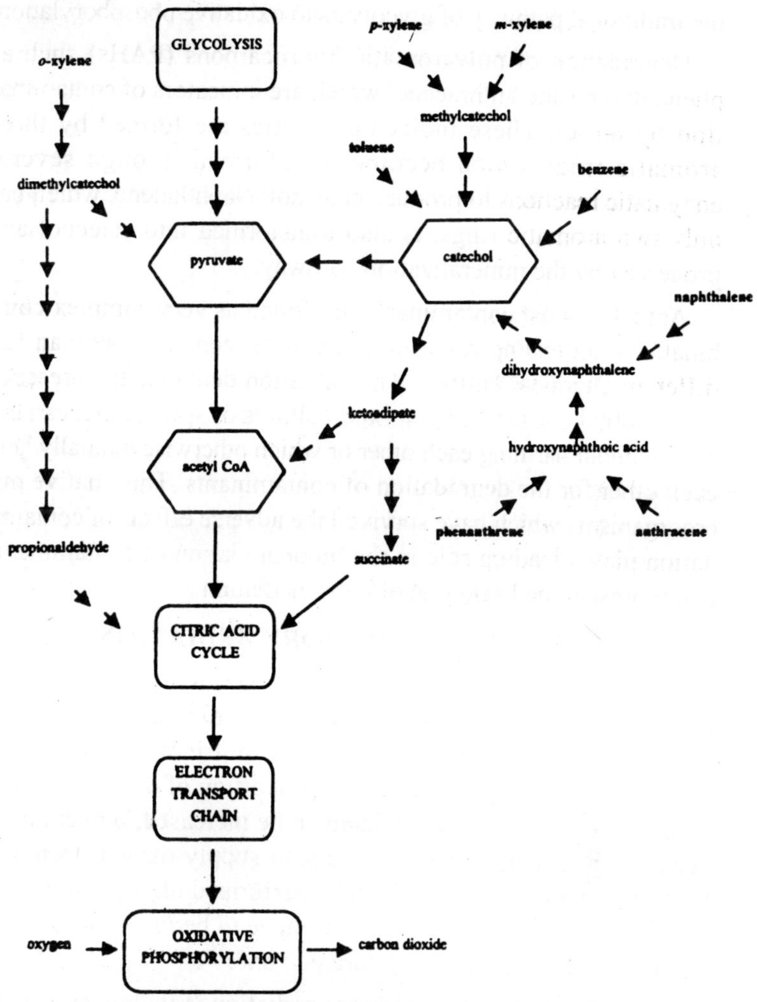

Figure 5.2 : ***Integration of degradative pathways for some contaminants.***

benzene, toluence and xylene isomers. An intermediary catechol is formed during degradation, which can be degraded through fermentation or respiration, by means of which it can reach other intermediaties of the classic pathway that leads to the generation of carbon dioxide, such as pyruvate or succinate. Summarized the integration of degradative pathways for some contaminants with the traditional pathway of glycolysis to oxidative phosphorylation.

Degradation of polyaromatic hydrocarbons (PAHs) such as phenanthrene and anthracene, which are indicators of contamination by diesel. These molecular entities are formed by three aromatic rings which become transformed through several enzymatic reactions to produce catechol. Naphthalene, which has only two aromatic rings, is also transformed into catechol and processed by the mineralization pathway.

Actually, most contaminants are found as very complex combinations; for example, gasoline and diesel contain more than 120 different chemical entities. This situation demands the presence of microbial consortia, i.e, mixed cultures of species that can co-exist without harming each other or which otherwise mutually help each other for the degradation of contaminants. Thise native microorganisms which have survived the adverse effects of contamination play a leading role in the biodegradation of contaminants and represent the backbone of bioremediation.

ADVANTAGES OF BIOREMEDIATION

Bioremediation is a versatile process because it can be adapted to suit the specific needs of each site. Biostimulation can be applied, but only when the addition of nutrients is necessary; bioaugmentation is used when the proportion of degradative microbial flora to contaminant needs to be increased; bioventing is needed when it becomes necessary to supply oxygen from air. Futhermore, bioremediation can be performed off-site when contamination is superficial, but it will have to be *in situ* when contaminants have reached the saturated zone.

One important feature of bioremediation is its low cost compared with other treatment technologies. According to Alper (1993), bioremediation is at least six times cheaper than incineration, and three times cheaper than confinement. It should however be mentioned that all cost comparisons cannot be generalized because they are only applicable to each particular case.

CONTAMINATED SITE

Not all contaminated sites are suitable for treatment with

bioremediation techniques; it will be necessary to demonstrate their efficacy, reliability and predictability in advance. To this objective site characterization should be performed to obtain information about three closely related aspects: the chemical nature of contamination, the geohydrochemical properties, and the biodegradation potential for the site.

- *Pollutant characterization.* It will be necessary to determine the composition, concentration, toxicity, bioavailability, solubility, sorption and volatilization of all pollutants.
- *Geohydrochemical characterization.* The physical and chemical properties of the geological material should be determined to be able to learn if the microenvironment is suitable for the biodegradative activity. In addition the geohydrological conditions of the site as well as direction and velocity direction of underground flow are of fundamental interest, particularly when contamination has reached the water table.
- *Microbiological characterization.* It is convenient to analyse the microbial flora in respect to degradative capacity and to the size of the native population with degradative potential.

Integration of the physicochemical and microbiological characterizations should correspond to the results of biofeasibility tests, from which it should be determined whether or not a certain biological treatment is applicable.

Once the site characterization is completed, it is important to proceed almost immediately with the activities leading to its clean-up because contaminants are not static. This is particularly true when pollutants are found in an aquifer.

The characterization of a contaminated site is of utmost importance because a better knowledge if it will facilitate the outlining of an *ad hoc* strategy for its bioremediation. The characterization should be performed in a logical sequence according to a previously established programme, to be able to respond to questions such:

- What chemical compounds are found as contaminants?
- Is the contamination superficial or has it affected the sub-soil?
- Are there any records to prove that the contaminants are bio-degradable?
- What is the depth and extension of the contaminant plume?
- What is the depth to the water table?
- Is the permeability of the geological material high or low?
- Are there microorganisms capable of degrading the contaminants?
- Is the environment suitable for microbial activity?
- Is it possible to 'build' a bioreactor at the site to be treated?

It answers are affirmative, then bioremediation could be applied, then it will be necessary to carry out biotreatability studies and the evaluation of a bench-scale or pilot-scale project, from which the full-scale process operation will finally be developed.

Something that is commonly encountered in practice is free contaminant in the aquifer; this must be removed before a bioremediation process is applied because the contaminants are toxic to the microorganisms. The latter are capable of tolerating certain concentrations, and some species show a higher tolerance than others, but it is not definit ively possible for microorganisms to develop within pure pollutants. This sort of detail should be taken into account when scale-up of the process is being outlined; otherwise a complete failure of bioremediation can be expected.

SUITABILITY OF THE SITE FOR BIOTREATABILITY TESTS

Once information has been gathered on the characteristics of the contaminated site it will be possible to identify its specific requirements. The fact of detecting contaminants at ground surface, in the aquifer or at the mid part of the unsaturated zone, will suggest a strategy of specific bioremediation for each particular case. Therefore, the biotreatability testing procedure will have ot be suited to each specific site.

Biotreatability tests are generally performed at a mesocosm level, trying to maintain the environmental conditions that will prevail during treatment in the field. If shallow strata are to be treated with an off-site process, large trays or jars to hold several pounds of soil could be used; it will therefore be necessary to keep humidity and homogeneity constant so that the microbial activity takes place in the whole volume of soil to be treated. If contamination is detected at the water table, columns packed with contaminated soils will be the preferred experimental model for biotreatability tests. It will be necessary in this case to determine the groundwater flow rate to be able to define the operating mechanism of the columns.

When characterization studies indicate that the microbial population with degradative potential is limited or practically zero, it will become important to add exogenous microorganisms. For very practical cases such as bioremediation two alternatives exist. The most common solution is to add commercial compounds; this way is more accessible and faster but it has the least likelihood of success. The other more, interesting, solution is to isolate the rather few degradative microorganisms that were obtained during site characterization and to promote their growth to obtain a culture that can be used for inoculation purposes. This method is safer although it has the shortcoming of requiring a longer time to increase the microbial biomass.

The purpose of biotreatability studies is to predict the behaviour of the process and to determine the nutritional requirements for microorganisms to perform biodegradation. The following measurements are thus required:

- Oxygen consumption
- Carbon dioxide generated
- Exhaustion of added nutrients (particularly nitrogen and phosphate sources)
- Contaminant removal

It is necessary to include biotic and abiotic controls to ensure that the contaminant is removed by a microbial activity. When tests are properly carried out it will be possible to predict the behaviour of bioremediation and the time for large-scale application.

For the biotreatability tests to actually represent the field conditions it will be necessary to adopt microenvironmental conditions as close as possible to those encountered at the site to be treated; otherwise the benefits obtained will be minimal.

Because biotreatability tests are time consuming and costly, when bioremediation is applied at a commercial scale these tests are not always performed; nutrients are incorporated empirically, based on past experience. The results are eventually satisfactory but most of the applications are bound to become a complete failure. This should be taken into account because a bioremediation failure may lead to further problems.

FROM LABORATORY TO FIELD

Even though techniques for growing microorganisms at a commercial level have been perfectly established, additional studies on technologies for bioremediation of soils and aquifers to promote, facilitate or expedite microbial activity in the field are needed. The more relevant aspects are related to bioavailability, the concept of bioreactor, the supply of oxygen, and mass transfer.

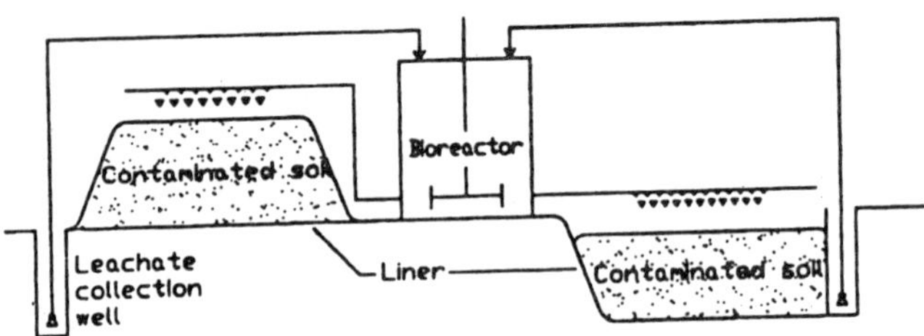

Figure 5.3 : ***Composite dioagram for off-site bioremediation techniques: left, a biopile; right, a biocell. Both have collection wells for leachate recirculation.***

Bioavailability

For the biotransformation reactions to be carried out it is necessary for the contaminants to be made ready for attack by the microorganisms. Enzymatic reactions generally occur in aqueous solution but when pollutants are insoluble in water – such as in the case of pertoleum hydrocarbons – their energy yielding will be slow due to surface tension between the equeous and the organic phases. To solve this problem special attention has been paid to the production of surfactants that could be incorporated into the medium and improve the bioavailability. A surfactant acts by 'solubilizing' organic contaminants in the aqueous medium, thus making them accessible to the enzymes responsible for their biodegradation. However, the uncontrolled use of surfactants courses many problems; these are refered to later.

The Concept of Bioreactor

As opposed to conventional biotechnological processes, in which previously built reactors are used, it is necessary for bioremediation purposes to 'build' the bioreactor at the contaminated site. When pollution covers shallow soil layers, it is recommended to excavate the material and to carry it somewhere else to build a biopile or a biocell. The difference between these two concepts resides in the fact that in the biopile the contaminated material is piled on the ground surface whereas for the biocell it is necessary to perform an excavation in a clean site to deposit the material for further treatment. It will be neccessary in both cases to place liners to confine the contaminated material and to prevent leachates from seeping toward the clean soil during treatment.

When contamination has reached the water table, the bioreactor can be built through the bore-holes. These are always drilled in even numbers; half are used for extraction and half for injection. Depth and location of wells is determined from geohydrological characterization of the site, from which the configuration of the contamination plume and the direction of the underground flow can be determined. The number of wells can be chosen once the radius of influence of each well has been established in terms of porosity and permeability of geological material, and the flow rate and direction of underground flow. The operating procedure of the wells determines the dimension and control of the bioreactor.

As is occurs in wastewater treatment procedures, it is not possible for bioremediation to work under sterile conditions. The diversity of the microbial population is set in terms of the kinetic characteristics of each of the species involved; it usually happens that native microorganisms that have developed a biodegradative capability in the same microenvironment where contaminants are present become the most widespread.

Oxygen Supply

Although the degradation of contaminant compounds under anaerobic conditions has been reported elsewhere, aerobic metabolism is the preferred choice to apply bioremediation techniques because reactions occur at a faster rate and complet mineralization so achieved.

When compounds used as a source of carbon lack oxygen in their molecule, such as in the case of hydrocarbons, oxygen demands for their biodegradation are higher than those required by oxidized compounds such as sugars.

Oxygen supply is one of the major engineering challenges for bioremediation applications in soils because a solid medium is encountered; in aquifers, it becomes more difficult to solubilized molecular oxygen as the depth of the subsoil increases. This has promoted the design of *ad hoc* aeration equipment and the search for alternatives to supply oxygen through highly oxidized inorganic compounds not related to alternate energy-yielding pathways so that full mineralization is achieved. As example of these oxidized compounds mention can be made of peroxides.

Mass Transfer

The concept of mass transfer in bioremediation basically refers to the homogeneity of the system – i.e. that in all of the points inside the bioreactor the same microenvironmental conditions exist to promote microbial activity. This includes nutrient concentration, humidity, pH, and concentration of oxygen available.

Bioremediation becomes more difficult when the geological material is basically a clay because its low permeability prevents mass transfer in the system. This is important when contamination has reached the water table and an *in-situ* treatment is available.

For superficial soils this problem can be overcome if sand or agroindustrial residues are added to increase the permeability.

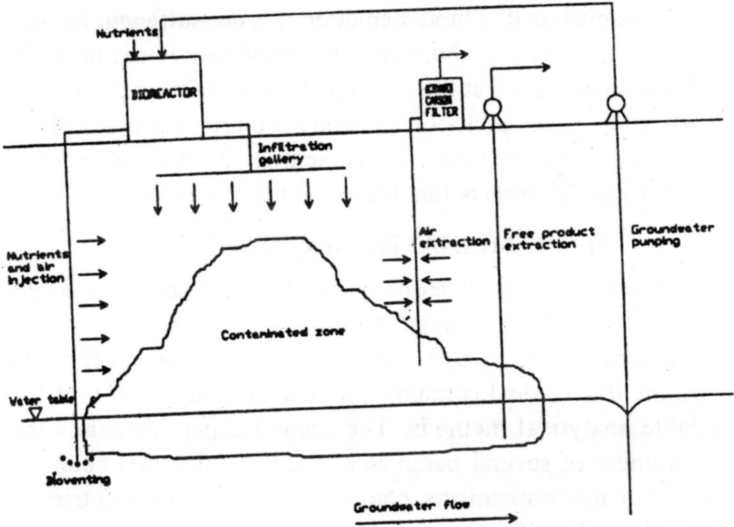

Figure 5.4 : ***Composite diagram for in-situ bioremediation techniques, including nutrient percolation, vapour andair extraction, bioventing, pump-treat-injection and free product recovery.***

In contaminated aquifers, the most popular auxiliary technique is the pump-treat-injection that involves the extraction of groundwater, its treatment at the surface, and its subsequent recharging into the aquifer. For bioremediation purposes treatment is carried out in a bioreactor where the degradative microorganisms may be confined; alternatively, the organisms can be recycled and leave the reactor at the surface to become reactivated. Although thus technique is widely used, there are certain aspects that determine its successful application; for instance, contaminants can be heavily adsorbed by geological material or may be present in low-permeability zones this restraining the mass transfer. In other cases, it becomes difficult to reach the clean-up levels required because the low concentrations of contaminants that microorganisms use as a substrate are not sufficient to support their microbial activity and they start to die. If this occurs, treatment becomes very costly because of the power requirements demanded by pumping.

If the previous concepts are taken into account, it will be possible to establish that in bioremediation it is not sufficient to work with ideal cases in the laboratory; it is important to maintain the conditions that prevail at the site and to use suitable experimental models so that the results of the study are representative of the scope intended for the field. This is the true objective of performing biotreatability tests before scaling-up is carried out it.

BIOREMEDIATION MONITORING

Monitoring of a bioremediation process is essential to determine two fundamental aspects: the degree of contaminant removal and the catabolic microenvironment. For the former it suffices to determine the residual contaminant concentration by any of the available analytical methods. The second aspect demands the measurement of several parameters such as microbial count of degradative microorganisms, concentration of residual nutrients, pH and humidity (for superficial geological materials), among others.

If the microenvironmental conditions where biodegradation is taking place are periodically determined, it will be possible to maintain each of the parameters within the levels in which the highest metabolic activity can be reached.

CLEAN TECHNOLOGY

The global market of bioremediation is becoming increasingly wider and more successful because it is regarded as a clean technology. This is mainly due to the fact that contaminants can be transformed into environmentally harmless compounds and some can be fully mineralized. It is also important because microorganisms die when there are no more pollutants to use as substrate.

Something that favours the image of bioremediation of that the soil, once the treatment is over and the degree of pollution very small, can be used for growing plants with the purpose of reintegrating it to its original biological functions.

On the other hand, it is a well known fact that many of the technological advances have been accompanied by environmental deterioration; if bioremediation is carelessly handled, it can lead to failures and to even worse environmental disasters. To keep the concept of bioremediation as a clean technology it will be necessary to perform a rigorous analysis based on ethical environmental concepts and on principles of sustainability that in many cases oppose the economic interests of commercial enterprises.

The increasing number of bioremediation companies on a world-wide scale is mainly due to the fact that this technology is economically feasible. These corporations offer not only environmental services but also related consumables that are basically microbial products, nutrients and commonly biodegradable surfactants. If these products are applied only as recommended, in the minimum necessary amounts and under treatment control, the technology will be indeed successful; otherwise, the surroundings of the site being treated can be adversely affected.

Some of the risks involved in a careless application of bioremediation can be described as follows:

- The uncontrolled addition of surfactants to aquifers can help the dispersion of contaminants rather than their full degradation.
- The excessive use of inorganic compounds used as nitrogen and phosphate sources that can be transported to lakes or lagoons favours the growth of undesirable species; an example is provided by eutrophication.
- Where a native microbial population with degradative capabilities exists, it is better to stimulate its activity *in situ* rather than applying exogenous microorganisms that will eventddually die from competition in the natural environment.

An additional aspect is the increasing interest to apply genetically engineered micro-organisms (GEMs) to expedite bioremediation. It is convenient in this case to mention that all the mechanisms that govern a natural environment such as an aquifer are not yet fully understood; furthermore, many factors are involved in the stability of the genetic information within the cells.

CONCLUSION

A large number of bioremediation technologies are now being developed and successfully implemented in countries that share certain environmental factors. However, when these technologies are transferred to countries with different environmental characteristics, the results are far from successul. It shall be considered in this respect that successful application of any type of technology depends on the need to perform studies for its implementation and innovation that could even result in further developments.

In the case of bioremediation technologies it should be understood that every soil has different characteristics and that no general rule exists for the microorganisms to readily adapt to any habitat. The soils in various parts of the world has distinctive physical, chemical and biological characteristics that make them different from each other.

Literature on new bioremediation technologies is being published every day; competition among big companies who have realized that bioremediation is a profitable money-making opportunity has also become evident. This competitiveness has been focused on the generation of increasingly efficient, but at the same time more sophisticated and expensive, technologies. The secret of a good bioremediation technology is to suit the know-how to every particular problem.

6

Polluted Soil

Soil-persistent compounds are identified as pollutants occurring as a result of industrial activities including fossil fuel exploitation, combustion and spills. Most common examples of such compounds are polycyclic aromatic hydrocarbons (PAHs) and polychlorinated biphenyls (PCBs), generally referred as non-ionic organic contaminants (NOCs). Since some of the NOCs are suspected to be hazardous or even human carcinogens, remediation studies and new technology development are now invoked as urgent needs. An interesting way of eliminating NOCs is bioremediation.

Bioremediation can be defined as the manipulation of living systems to bring about desired chemical and physical changes in a confined and regulated environment. Bioremediation uses micro-organisms (bacteria, yeast or fungi) of microbial processes to detoxify and degrade environmental contaminants rather than the conventional approach of disposal. As bioremediation is strongly limited by biodegradation, a serious distinction between the concepts of biotransformation and complete biodegradation or mineralization is required. In this chapter, the term 'biotransformation' means any transformation of the structure of a compound by living organisms or enzymes, while 'biodegradation' involves complete breakdown or mineralization of molecules to carbon dioxide and water. For example, Cerniglia *et al*. (1994) found that the filamentous marine fungus *Cunninghamella elegans* biotransforms the PAH benz[a]anthracene to *trans*-dihydrodiols and related compounds, which accumulate, but no mineralization was observed. Therefore, while biodegradation assures the elimination of contaminants, biotransformation does not.

Several factors can limit the rate of biodegradation of soil contaminants, including temperature, pH, oxygen content and availability, soil nutrient and availability, soil moisture content and the physical properties of contaminants. In addition, NOCs have low aqueous solubilities, low dissolution rates and they are, as soil contaminants, strongly bound to or adsorbed onto solids. The biodegradation of such compounds in the natural environment may be restricted, mainly because: (i) the native population of microorganisms is absent or extremely poor and bioreaction limits the process, (ii) the compounds to be degraded are not available to microorganisms and mass transfer becomes limiting, and (iii) compounds or their metabolic intermediates are toxic to the native microorganisms and biodegradantion is inhibited. The lack of suitable microorganisms can be overcome by exogenous inoculation, but distinction between availability due to mass transfer when residual fraction remains undergraded and accumulates in historically contaminated (i.e. aged) soils.

Bioavailability can be defined as the availability of chemicals to potentially degradative microorganisms. If pollutants become unavailable it is because the rate of mass tranfer is zero – as in the case of bound residues in aged soils. Therefore, bioavailability cannot be regarded as a property of one chemical molecule by itself but a property of the complex molecule–environment (e.g. soil moisture, salinity, aqueous phase pH, etc,).

Recent studies of PAH biodegradation, ranging from the structurally most simple molecules, to the more complex, agreed that bioavailability was the main constraint to overcome and a relevant question to be answered before palnning any bioremediation strategy.

The aims of this chapter are to elucidate the importance of bioavailability compared with other factors also involved in soil bioremediation and to discuss constraints and alternatives to increase bioavailability of recalcitrant compounds found in contaminated soils.

BIOREMEDIATION:TECHNOLOGY

Bioremediation of contaminated soils has been used as a safe, reliable, cost-effective and environmentally friendly method for degrading various NOCs, from oil, diesel and gasoline, PCBs cases, a common final result prevails: a substantial portion of the pollutant remains unaltered or only partially altered in the bioremediated soil, and target NOCs build up. When incomplete biodegradation is found, it results in accumulation. The reasons why certain pollutants should accumulate in the site are not quite clear. Huesemann (1997) suggests that incomplete NOC biodegradation is caused not only by bioavailability limitations but also by the inherent resistance of certain compounds to break down in the evance of the differences in these two basic concepts is in bioremediation efforts: if biodegradation is incomplete due to bioavailability constraints, one can limit remediation effort because accumulated compounds should be trapped into soil matrix in a non-bioavailable and thus safe form; on the other hand, if some toxins accumulate much more effort must be made. Mistakes underestimating toxicity must be avoided. In other words, adequate attention must be paid to the toxicity of incompletely biodegraded as well as to the biotransformated, still bioavailable, by-products. To understand accumulation, a reasonable explanation is suggested in an interesting critical review: sorption (and especially desorption) in natural particles can be extremely slow, which could partially account for the observed difficulties of accumulation. True sorption equilibrium is reached in months or longer, while in the laboratory the time scale is reduced to weeks or even hours. To avoid misunderstanding and erroneous conclusions, two different experimental domains must be distinguished: (i) fast sorption, and (ii) slow sorption. Fast sorption lacks long-term kinetic studies and results are not necessarily conclusive. In the slow sorption domain, great effort is needed to define the origin and constraints of slow desorption for remediation purposes. The list of constraints to study includes composition and nature of the contaminant, temperature climatic considerations, bioavailability and multiphase partitioning (non-aqueous and solid), presence of toxic compounds (the contaminant itself, salinity, heavy metals) and soil texture and structure.

Unfortunately, the vast majority of published work describes experimental techniques in which recalcitrant molecules were added to samples (spiked samples) of soil, sediment or any other artificial medium, falling in the fast sorption domain. Very few studies describe the biodegradation of such compounds on soils that have been contaminated for years.

In conclusion, despite the potential, bioremediation is considered an emerging technology because final effects (e.g. accumulation) are not well understood and are unpredictable. It is possible to surmise the fate of both the biotransformed molecules and exogenous microorganisms, but final consequences in the bioremediated site are uncertain; strict monitoring must be assessed over-the next decades and much more predictive research must be done. Finally, the concepts of bioavailability and soil bioremediation are strongly related. Moreover, it is possible to state that soil bioremediattion proceeds if bioavailability has been successfully overcome. To assure that bioavailability is not a limitation it is necessary to identify the influencing factors that should be eventually improved.

BIO CONSTRAINTS

A number of constraints that could influence bioavailability must be identified before planing a bioremediation strategy. These include (i) the nature of microorganisms, (ii) the composition and properties of pollutants and (iii) the nature of soils.

The influence of the nature of the microorganisms used was studied. They studied the biodegradation of naphthalene, a minor component of refined petroleum, considered as a model because its moderate solubility (31.7 mg l^{-1}) and susceptibility to degradation by a large number of microorganisms. The authors examined the rates and extents of degradation in soil-free and soil-containing systems in a comparison of two bacterial species: *Pseudomona putida* ATCC 17484 and a Gram-negative soil isolate, NP-AlK. In the first case, both the rates and extents of naphthalene desorption from soil (i.e. naphthalene was relatively unavailable). They concluded that there are important organism-specific properties that make it difficult to generalizedd the concept of bioavailability of soil-sorbed substrates.

Deschenes *et al.* (1996) Demonstrated that the biodegradation of the three-ring PAH found in an aged soil were more readily mineralized than the four-ring PAH series. The authors observed no biodegradation at all for PAH having more than four rings. These results confirmed a general rule: the higher the number of fused benzene rings, the more resistant the contaminant to breakdown. Johge *et al* (1997) Studied the effect of fuel oil composition on bioavailability in a lysimeter and in laboratory scale. They found that two different mechanisms, depending on the *n*-alkane (*n*-C16 to *n*-C20) ratio, controlled the bioavailability. At concentrations higher than 4.0 g kg^{-1}, bioavailability was controlled by solubilization from the non-aqueous liquid phase to the aqueous soil water phase but below this concentration desorption and diffusion became rate limiting. An increase in biodegradation rates with decreasing carbon number was also observed. Finally, they proposed that monitoring *n*-alkane ratios during the course of treatment may be important in identifying bioavailability restrictions in aged soils.

Another interesting approach to elucidate the effect of the nature of contaminants to bioavailability was reported by Sugiura *et al.* These authors studied the biodegradation by different bacteria of four crude oil samples from which the readily biodegradable fraction had been removed. They found that the same compounds in different crude oil samples were degraded to different extents by the same microorganism. One possible explanation for this was the bioavailability of the monitored alkanes, which were different in different crude oil samples.

A third bioavailability constraint is soil composition. Because it regulates (i) the maximum attainable capacity (partition coefficients) of pollutants in the liquid and solid phase (ii) the kinetic and desorption rate of chemicals and (ii) the aggregation level, grain size and the porosity of particles in which contaminants are trapped. Soil texture also influences the water regime (e.g. infiltration), gas exchange (producing or avoiding anaerobic cores), local temperature, effective diffusion (which depends on carbon content of matrix soil) and micropore tortuosity.

A number of mathematical modelling studies followed bioavailability constraints such as the nature and extent of contaminants, and the nature of soils, simultaneously. In models a set of hypothetical considerations must be made and validated. The final objectives of modelling are to predict transport behaviour of NOCs and discriminate between transport and reaction limitations, and ultimate condition of understanding bioavailability as a transport phenomenon. Bioavailability has beeen modelled in the traditional chemical reaction engineering way – i.e. through energy and mass balances coupled to reactions in which the complex soil–contaminant–microorganism is assumed to be the catalytic system. For example, a kind of Thiele module has been developed involving diffusion coefficient, biodegradation rate coefficient, soil aggregate radius and adsorption capacity, which may serve as criteria to assess whether intraparticle diffusion resistance can be ignored. Chemical engineers have devised and interesting generic mathematical concept named bioavailability number (Bn), which resembles the inverse of the dimensionless Damkoller number. Calculation of Bn can give an idea of the local importance of mass transfer at values less than, unity. In general, mathematical models fit well to experimental results when a single contaminant is considered (for example phenanthrene) in an aqueous clean solution (sterile medium pure inoculum), where such a molecule is always bioavailable. However, models fail when complex mixtures of contaminants or soil are present.

In conclusion, the nature of the microorganisms could influence the final results; the bioavailability is not solely determined by the composition and chemical structure of pollutants but other factors such as soil matrix. Until recently, published reports on degradation kinetics of several simultaneously present NOCs, as those win aged soils, have not been available and further research in this direction is expected. Increasing local bioavailability is a serious challenge to overcome in future research.

BIOAVAILABILITY

Three general strategies to increase bioavailability have been developed: (i) use of biosurfactants, (ii) addition of organic solvents and (ii) addition of synthetic surfactants. Owing to its importance and potential of application, the addition of synthetic surfactants is separately discussed in a following section.

Plants, animals and microorganisms can naturally produce surfactants; those produced by microorganisms are known as biosurfactants. The principal property of biosurfactants is their capacity to reduce the surface tension of liquid media. Zhanganl Miller (1992). Postulated that if surface tension is reduced it results in an increase of aqueous dispersion of NOCs, consequently solubility and rates of dissolution are also increased. Once dispersion is increased, bioavailability and likely biodegradation should be enhanced.

Biosurfactants are exopolymers (e.g. polysaccharides, gums) mainly produced by bacteria under growth-limiting conditions (e.g. high carbon to nitrogen ratios and iron limitation) or when poorly soluble substrates such as *n*-alkanes are present. Among the most studied biosurfactants are the rhamnolipids produced by *pseudomanas aeruginasa* the use of biosurgactants is attractive because they are natural products and therefore biodegradable, showed that purified rhamnolipids produced by *P. aeruginosa* UG2 was readily biodegraded by native PAH degraders found in aged creosote-contaminated soil. These results suggest that two different, and not necessarily coupled microbial activities must be distinguished: (i) biosurfactants produced under limited conditions by bacterial and (ii) biodegradation of contaminants as a result of increased bioavailability. As microbial growth on hydrocarbons has been associated with the production of biosurfactants, isolated bacteria with two of the desired characteristics: PAH metabolizing bacteria capable of producing biosurfactants. They found that in the presence of naphthalene the bacteria produced biosurfactants which promoted the solubility of naphthalene. The authors suggest that if microorganisms promote the solubility of their own substrate, biosurfactants are naturally produced as a part of their strategy for growing on such substrates where bioavailability is essentially reduced. Apparently, the potential advantage of using biosurfactants requires presence of the biosurfactant producer rather than addition of the isolated purified biosurfactant.

Addition of organic solvents changes the polarity of the soil–water environments, influencing partition and consequently bioavailability of non-ionic pollutants. Furthermore, appropriate organic solvents disperse NOCs up to molecular level, promoting a thin layer interface to which NOC degraders could adhre. Despite the potential of this bioremediation principle few research groups have paid attention to the effect of solvent addition. Conducted experiments in liquid cultures with an *Arthrobacter* strain and demonstrated that the extent of biodegradation of naphthalene dissolved in heptamethylnonane was increased by addition of the solvent. Biodegradation was proportional to the volume of solvent and attributed to bacteria attached to the solvent–water interface. The authors concluded that adherence of cells to a slovent–water interface was a prerequisite for naphthalene utilization. Moreover, when a surfactant (Triton X-100) was added to prevent adherence of cells to the interface, biodegradation was also prevented. More recently, in and extensive work, used different strategies to increase bioavailability and likely biodegradation of pyrene. They studied the effects of non-ionic surfactants, hydrophilic (polyethylene glycol) and hydrophobic (heptamethylnonane, decalin, phenyldecane and diphenylmethane) solvents, which were inhibitory in all cases. They also experimented with the addition of paraffin oil. squalene, squalane, tridecyclihexane and tricosene: all increased pyrene biodegradation without being used themselves. Their explanation of enhanced biodegradation agrees with that proposed – that solvents promote adherence of organisms to the interface in the neighbourhood of dispered NOCs.

In conclusion, the production of biosurfactants can be considered as part of the metabolism of indigenous bacteria in contaminated sites. Microorganisms which produce biosurfactants could create a favourable local environment to enhance bioavailability of the NOCs attached to soil. Because this phenomenon does not assure biodegradation, future research looking for microorganisms and operation conditions that couple both the increased bioavailability and biodegradation is expected. With respect to the addition of solvents to increase bioavailability, the promising results discussed above now constitute a serious prospect for application and appear to justify further studies.

SYNTHETIC SURFACTANTS

The uses of synthetic surfactants to promote bioavailability of contaminant molecules attached to soil have recently gained attention, in particular, when insoluble persistent contaminants such as PAHs and PCBs in soils and sediments are present. Investigated the effects of four non-ionic surfactants on the bioavailability and rates of biodegradation of crystalline naphthalene and phenanthrene in agitated flasks. They found that the presence of synthetic surfactants increased both the apparent solubility and the maximal rates of dissolution. The rates of biodegradation of naphthalene and phenanthrene were also enhanced by He addition of surfactants in the dissolution-limited growth phase, indicating that the dissolution rates were higher than in the absence of surfactants. These authors also studied the toxicity of the surfactants, and observed no toxic effects due to the presence of up to 10 g l^{-1} of surfactants. Similar results showed that mineralization of naphthalene and phenanthrene was unaffected by the addition of non-ionic surfactants even above their critical micelle concentration (CMC). When solubilized be micelles of surfactants in aqueous phase, naphthalene or phenanthrene became bioavailable and degradable by microorganism. However, found that at concentrations above the CMC the mineralization of phenanthrene in soil–water systems was inhibited. Similar results were reported with *Mycobacterium* sp., in which the addition of non-ionic Trition X-100 below the CMC increased pyrene biodegradation but above the CMC severely inhibited mineralization. Furthermore, it was suggested that non-ionic surfactants at low concentrations may promote the biodegradation of adsorbed aromatic compounds in polluted soils, even when surfactant-induced desorption was not appreciable. Thibault *et al* (1996) conducted experiments with pyrene in spiked and soil-contaminated samples. They tested four synthetic surfactants, under saturated (soil slurries) and unsaturated conditions. The effectiveness of the surfactants in pyrene desorption increased with increasing concentration. Witconol SN70, a non-ionic ethoxylated alcohol, was the most effective surfactant for pyrene mineralization at unsaturated conditions but inhibited the action of microorganisms at saturated conditions, suggesting that excess of water plays an important role. Researchers need to take account of this in bioavailability and likely biodegradation studies.

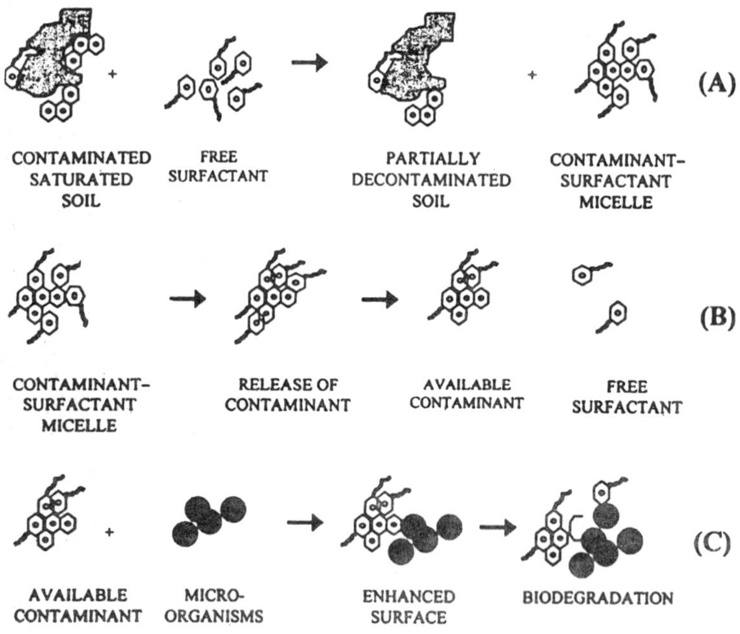

Figure 6.1 : ***Effect of surfactants onbioavailability. (A) Desorption of attached contaminant and micelle formation; (B) release of contaminants trapped into the micelles to the aqueous phase; (C) promotion of enhanced surfaces to facilitate biodegradation.***

Deschenes et al (1996).Demonstrated that the addition of sodium dodecyl sulphate (SDS) to creosote-contaminated soil increased the solubility of PAHs but that PAH biodegradation was inhibited, in agreement with the work of Tiehm (1994). Tiehm studied biodegradation in blends of PAHs by pure and mixed cultures in the presence of synthetic non-ionic surfactants and SDS; results indicated that surfactants increased the solubilization of SDS but that its degradation was inhibited. Non-ionic surfactants increased bioavailability and enhanced biodegradation of fluorene, phenanthrene, anthracene, fluoranthene, and pyrene, but differential toxicity was observed. Tiehm concluded that toxicity correlates with structural form: decreasing with increasing length of the surfactant.

To account for the inhibition of SDS degradation in the two previous cases, the same explanation was given: SDS was a preferential growth substrated in liquid cultured as well as in microcosm where 75% of initial SDS was biodegraded in 15 days.

Hydrophobic (cyclic ring) and hydrophilic (aliphatic side-chain) moieties illustrate the two functional components of surfactants. As general rule, the presence of a surfactant may help to enhance at least one of the following mechanisms (i) Desorbing contaminants from soul to the aqueous phase by releasing a complex contaminant-surfactant, which constitutes a pseudophase called *micelles*, plus partially decontaminated soil. (ii) Increasing the concentration of the hydrophobic compound in the aqueous phase by solubilization into micelles. At a CMC of the surfactant, colloidal aggregates are formed, providing increased solubilization or emulsification of the hydrophobic compound, (iii) Enhancing surfaces to facilitate transport of hydrophobic compounds from the micelle to the aqueous phase, releasing contaminant in available form plus free surfactant. (iv) Enhancing adherence between cells and substrate. Free surfactant could be assimilated and available contaminant might be attacked or removed by microorganisms.

In conclusion, the effectiveness of a synthetic surfactant in increasing bioavailability of soil-bound molecules depends on the nature of the microorganisms and contaminants, the type of surfactant and soil composition. Experimental results on the effects of surfactant addition on biodegradation of NOCs are not consistent and sometimes contradictory. Apparently, the observed positive effect of non-ionic surfactants in aqueous solutions is absent in the presence of contaminated soil. There is a lack of general explanations on the nature of the physical, physiological or biochemical phenomena involved. Although surfactants have been studied in complex water–soil systems, the mechanisms and effects are not well understood and more research in this direction must be done.

GENERAL RECOMMENDATIONS

An appropriate study of surfactant addition to increase bioavailaility should include the following separate studies:

- Surfactant to soil sorption kinetics must be assessed. There is a remarkable difference between studies based on clean and contaminated (e.g. aged) soil samples. illustrates the comparison: in a clean soil, a higher consumption of synthetic surfactant is required that in a contaminated soil. If some kind of binding force between surfactant and soil matrix (saturated and unsaturated) is observed, then the nature and extent of such bound will determine the role and effectiveness of the surfactant.
- Efficiency of NOCs extraction with and without surfactants must be measured. The presence of the surfactant should not interfere with the NOC extraction methods.

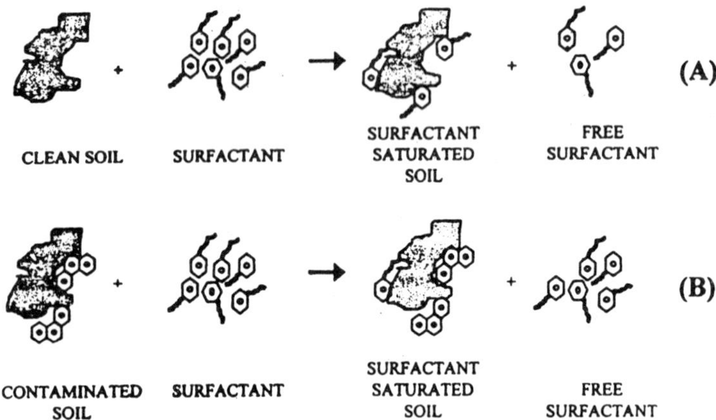

Figure 6.2 : ***Saturation of clean (A) and contaminated (B) soil by a synthetic surfactant.***

- Presence of the surfactant should not disperse soil samples. Dispersion of soil aggregates will result in an altered form quite different to that found on site.
- Partition coefficient and CMC must be evaluated. Use of culture medium, if any, instead of pure water is also sug-

gested. An effective surfactant will produce higher partition and lower CMC values.

- Intrinsic toxicity (e.g. altering cell membrane integrity) of the surfactant must be studied. For example the assimilation of conventional carbon sources such as glucose with and without surfactant can give an idea of the extent of intrinsic toxicity.
- Intrinsic biodegradability of the surfactant must be studied. A careful study regarding the ability of selected microorganisms as NOC degraders to either eliminate of biotransform surfactants is suggested. In some cases, surfactants could be used as preferential source of carbon, this a compromise between total biodegradability, which avoids surfactant accumulation, and preferential source of carbon must be found.

CONCLUSIONS

1. To achieve an appropriate soil bioremediation approach, a formal distinction between the concept of biotransformation and complete biodegradation must be envisaged.
2. Regarding the role of bioavailability of contaminants in aged soils, spiked samples studies do not allow scale-up of conclusions to the field level.
3. Low bioavailability of hydrophobic compounds bound to soil matrix is now recognized as a major bioremediation constraint to overcome. In most of the cases, mass transfer limitation prevented the success of biodegradation.
4. Mathematical conceptualization of bioavailiability allows estimation of such significant factors as to what extent biodegradation is limited by inherent microbial activity in comparsion to mass transfer constraints e.g desorption, transport throughout the solid matrix and dissolution rates.
5. Selection of an appropriate strategy to increase bioavailability strongly depends on concentration of contaminants in soils. Two extreme cases are: (i) low NOC concentration, which probably can be solved by using biosurfactants, and (ii) high

NOC concentration by the addition of organic solvents or related additives. However, both alternatives seem to be far from practical.

6. Among the different strategies to increase bioavailability, surfactant addition seems to be the most promising and reliable.
7. Surfactant addition to increase bioavailability involves a complex experimental task. In order to minimize risk and blurred conclusions, a number of parallel studies must be carried out.

7

Bioinsecticides

The human population is facing a great challenge to produce more food at a faster rate but in a sustainable way. In our discussion here, 'sustainable' means to have methods of production that protect the environment and the biodiversity and also keep natural resource for future generations. In this context, it is important to analyse the international situation: the world population keeps growing and it is estimated that the mark of 8000 million people will be reached by the year 2005. Most of the increase, 2500 million, will occure in developing countries, and nearly 85 % of the total will be living there. In order to cover the food needs, agricultural production will have to be doubled around year 2025; however hunger will still leave about one billion malnourished people.

During 1996 in Mexico the situation of food production was as follows: we imported 10 million tonnes of basic grains, 7 million of which were corn and 250, 000 beans, which represents an expenditure of $3 billion. This is a worrying prospect, not only due to the lack of rain in many agricultural sectors of the country but also because it indicates structural problems in the agriculture. In 1994 the FAO constructed a simulation model to estimate domestic food demand to the year 2010 (FAO. 1994); the model takes into account all uses of agricultural products (direct human food, industrial uses, animal feed, etc.) and considers that the NGP will grow at 4.8% while population will increase at 1.9% between 1989 and 2000 and 1.6% in the period 2000–2010 In these conditions the total internal demand for agricultural products will grow annually 2.6% in the period 1989–2010, which represents an annual increase of 0.4% of calories available for the population. We can anticipate today that these estimations will never occur, because the growth of the NGP will be smaller than estimated. Even though the FAO's model indicates that future growth of demand will be smaller than in the last three decades, the demand will grow faster than the domestic production,

TABLE 7.1 : FUTURE DEMAND FOR FOOD PRODUCT (THOUSAND T/YEAR)

Product	1988–1990	2010	Increase
Rice	786	1 349	563
Corn	16 395	23 675	7 280
Wheat	4 731	7 613	2 882
Beans	1 384	1 523	139
Sorghum	7 851	12 582	4 731
Vegetable oils	1 248	2 414	1 166
Milk	8 988	13 759	4 771
Bovine meat	1 931	3 129	1 198
Eggs	1 058	1 727	669
Pork meat	835	1 750	915
Bird meat	768	1 793	1 025

For these reasons the only way to cover the food demand will be through food imports. The yearly rates of imports will be high: 6.7 % for wheat, 5.2 % for rice and 3.9 % for corn. The agricultural commercial deficit will grow 57 and 49 % for the periods 1989–2000 and 2000–2010, respectively.

For sevçral years domestic production has been increased by using new agricultural methods developed by adoption of technology of the type known as 'Green Revolution' and also by the opening of large areas for irrigation, which at the same time had important negative environmental impacts. The contamination related to agricultural inputs has been growing mainly due to the extended use of pesticides. These products have grown 5% annually, from 14 000 t in 1960 to 60 000 t in 1986, as a result of the diffusion of this type of technology and also due to the fact that many pathogenic agents have developed resistance to chemical products and new pests have appeared. From 1988 to 1992 the import of plaguicides to Mexico increased from 30 000 t to 60 000 t, but how much of this is applied domestically, and the amount of pesticides produced and applied in Mexico, is not known.

TABLE 7.2 : TRENDS, FUTURE DEMAND AND IMPORTATION OF FOODS PRODUCTS: ANNUAL GROWTH RATE.

Product	Domestic demand Historical trends		Projection 1988–2010	Importation 1988–201
	1961–1990	1980–1990		
Wheat	5.1	2.1	2.3	6.7
Corn	3.4	2.7	1.8	3.9
Rice	3.6	3.0	2.6	5.2
Sorghum	11.2	0.0	2.3	2.3
Cereals	4.8	1.8	2.1	4.0
Meats	5.3	3.5	3.1	0.9
Total foods	4.8	2.6	2.7	3.0
Total (foods/other products)	4.6	2.6	2.6	3.0

It is thus necessary to increase food production but how could it be done keeping the environment and the biodiversity safe and conserving the natural resources for the future? A satisfactory answer faces problems of large technological complexity: agricultural productivity must increase but at the same time consumption of agrochemicals should decrease. This goal can be reached at least partially through

- replacing traditional agrochemicals (insecticides, fungicides, herbicides, etc.) by novel and less toxic products, susceptible to short-term biodegradation in the environment;
- national policies to reduce the input of agrochemicals;
- support of research and development directed towards integrated pest management.

In this chapter, we will describe some of the most interesting aspects of the most widely commercial used bionisecticide, *Bacillus thuringiensis* (*Bt*), because it is a good model to understand the problems of replacing toxic agrochemicals with biodegradable products. The areas that will be covered are:

- the search for novel *Bt* activities against different kind of insects;
- study of the mechanism of action of the *Bt* δ-endotoxins *Bt* genes (*cry* genes)
- development of new systems for use of *Bt* toxins.

BIOINSECTICIDES BASED ON *BT*

Bacillus thuringiensis is a Gram-positive soil bacterium that produces insecticidal crystal proteins (ICP), also known as δ-endotoxins, during the sporulation process; these are toxic to the larvae of a number of destructive insect pests in forestry and agriculture. The spores and the crystals produced by this bacterium were discovered at the beginning of the twentieth century, and commercial products based on it were used until 1956 in France. However, the δ-endotoxins (also called cry proteins), the main constituent of the crystals, have recently attracted the attention of several research groups and also of some international companies for two reason: δ-endotoxins with novel activities have been discovered, and the arrival of molecular biology to agriculture has generated transgenic plants containing the *cry* genes.

In 1990 it was reported that *Bt* represents around 95 % of the total biopesticides market, its sales volume being $260 million. It has been estimated that the demand for these products will grow at least 10 % annually until the end of the twentieth century. The principal producers of *Bt* biopesticides are Abbott, Sandoz, Du

Pont and Novo Nordisk. The market distribution by regions is: USA and Canada 50 %.

It we follow the historical development of the discovery of δ-endotoxins with novel activities we can see that its growth rate has been practically exponential since the 1980s and today we can foresee that with new biological methodologies available more toxins will be found with novel activities. Particularly it must be pointed out that *Bt* toxins are not toxic to humans and other mammals after 40 years of use and this characteristic has made them very attractive because it is not necessary to carry out long and costly clinical trials to probe its lack of toxicity and, at the same time, the past experience of commercial use makes easier and faster the introduction of similar products.

At the present time several *Bt* stains have been identified with different insecticidal activities; most of them are toxic to lepidopteran larvae (caterpillars), but there are also strains with toxic activity agains dipterans (flies and mosquitoes), coleopterans (beetles), mites and ants (hymenopterans), platyhelminthes (flat worms), protozoans (amoebae) and nematodes (roundworms). It is estimated that the list of susceptible organisms will keep growing and this area, the search of toxins with novel activities, is the principal goal of many research groups around the world, in academia as well as in private companies.

Mechanism of Action

The elucidation of the mechanism of action of the *Bt* toxins is, besides the search for novel activities, the big area of interest in the field of the *Bt* bioinsecticides. This knowledge will allow the design of more potent molecules or manufacture of molecules with a novel and/or higher spectrum of action. The current knowledge about the mode of action of the *Bt* δ-endotoxins is illustrated in.

It is known that cry proteins accumulate within the cell, forming crystals during the sporulation process. In the crystals, the proteins are immature products (protoxins); in order to be active the crystal must be solubilized in the extreme pH conditions of the insect midgut and in the same place also must be activated by proteolytic processing. The mature products or toxins bind to specific receptors in the brush border membrane of the midgut columnar cells where, after a substantial conformational change caused by interaction with the receptor, it is inserted into the membrane and, presumably by oligomerization, forms pores. These pores disturb the endogenous permeability properties of the target membrane. After that, there is a massive entry of water into the cells, causing the destruction of the epithelial tissue and finally the death of the larvae.

The putative receptor has recently been isolated and characterized. It has been reported that the Cryl Ac receptor in *Manduca sexta, Heliothis virescens* and *Lymantria disparis* is an aminopeptidase N that is linked to the cell membrane by a glycosyl–phosphatidylinositol anchor.

Using several methods *in vitro* to know with more detail the effect of the δ-endotoxins in the membrane that contains the receptor, several authors have suggested that the primary action of the cry proteins in the membrane is to induce some endogenous proteins to allow the flux of ions through the membrane – in other words, to open an ionic channel selective to cations, or forms (alone or with the receptor) a cationic channel. Some δ-endotoxins have been studied incorporating them into planar bilayers formed by synthetic lipids, where it has been shown that these toxins could form ionic channels *per se*. The ionic channels formed by these proteins allow the perferential flux of monovalent cations; other important characteristic of these channels is that apparently they are formed by one or more molecules, vecause its conductance (the property that describes how much resistance must an ion overcome to pass through the channel) is very variable (500–4000 pS). It is important to point out that in order to observe these channels experimentally micromolar concentration of toxins have to be use, while the effect of there in the apical membrane of the epithelial cells occurat nanomolar of picomolar concentration. This fact suggests that the receptor reduces the toxin concentration required for the protein or the protein–receptor complex to insert and form pores into the membrane. There is still an open question: do the channels observed in the black lipid bilayers in the absence of the receptor correspond to the channels formed by the cry proteins *in vivo*?

At least 60 different *cry* genes have been described and classified considering the similarity of the amino acid sequence in 19 different groups (1, 2, . . . , 19) and subgroups the cry protein toxic regions it has been possible to identify five very conserved regions (blocks 1, 2, 3, 4 and 5) that are separated from others of low similarity and variable length.

Functional Significance

The tridimensional structure determined by X-ray diffraction studied of the toxic portion of one of these proteins, Cry3A, has shown that it is organized in three domains, to each of which a specific function has been assigned. Domain I (residues 1 to 290) is constituted of six amphipathic helices surrounding another hydrophobic one, α-helix 5; these are possible constituents of a structure that forms pores. Domain II (residues 290–500) is formed by three β-antiparalled layers that end in loops in the vertex of the molecule; this conformation is called β-prisms. The region that forms the loops is the less conserved (hypervariable) region; the interchange of fragments of the hypervariable region between three closely related proteins allows interchange of specificities, and for this reason the domain is associated with the interaction with the receptor. Domain III is composed of δ-sheet arrays in the form of a β-sandwich (this topology is called double helix β); toxins with mutations in this domain are very unstable to protease treatment, and the role of this domain is to be responsible for the five conserved regions are in the central part of the molecule, and it is proposed that the cry proteins have a common conformation and similar mechanism of action. This proposal was corroborated recently when the structure of another member of the cry family Cry 1 Aa, was published. There is only one significant difference between Cry 3 A and Cry 1 Aa: the loops of domain II of the latter toxin are longer. This structural information supports the idea that the hypervariable region is a determinant of the different specificity of each one of these toxins; Figure 7.1.

There is evidence that Domain I, specifically block 1 (this region corresponds to the α-helix 5), is involved in pore formation. Mutants of Cry 1Ac in α-helix 5 were equally effective as the

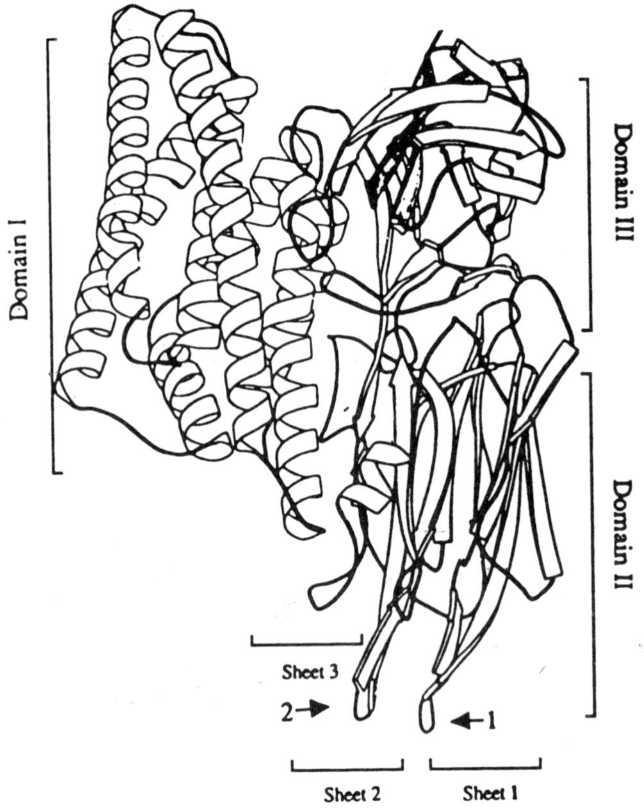

Figure 7.1 : *The three dimensional structure of* Bacillus thuringienisis *Cryt3a toxin. Domain I contains seven α-helixes, six of them arranged the seventh around, α-helix 5. Dopmain II is formed by three β-sheets (loops 1 and 2 are denoted by arrows). Domain III has a β-sandwich structure with a topology typical of double β-helix.*

wild type when binding to the receptor in bursh border membrane vesicles (BBMV) obtained from the midgut of three susceptible insects (*Manduca sexta, Heliothis virescens* and *Trichoplusia ni*); however, the activity in bioassays of some of these mutants decreased dramatically (between 10 and 1000 times) when proline or other charged residues were introduced. It was proved that these mutants lose the capacity to inhibit the leucine transport dependent on the K^+ gradient in the BBMV; this is an indirect measure that indicates that these proteins are capable of altering the vesicle membrane permeability. Other results that support the hypothesis that the central helix is a structural component of the pores were obtained when peptides corresponding to the α-helix 5 of Cry3A and Cry1Ac were synthesized; both peptides maintain the capacity of the whole molecule to form ionic channels selective to cations in black liped bilayers. In the literature it is proposed that at least one α–helix of Domain I (α-helix 7) can also participate as an structural component of the pore. This possibility has not been eliminated, although the evidence available suggests that α-helix 7 could participate in addition to the β-1 of Domain II as a bridge of communication between these two domains to allow the molecule to make the conformational change neccesary for the toxin to pass from a soluble conformation to a state that will insert into a hydrophobic environment, which is present in biological membranes.

During experiments of site-directed mutagenesis, where one or more amino acid residues present in the protein structure are substituted by others, it has been confirmed that the regions determinant of the high specificity of each one of the toxins are those corresponding to the loops in the apex of the molecule (Domain II). When this region in proteins Cry1Aa, Cry1Ab, Cry1C and Cry3A was mutated, it was observed that the affinity of the mutant proteins to the receptor are substantially changed.

Three of the five regions of conserved amino acids in the cry family are localized in Domain III. The evidence obtained from several authors suggests that preservation of the molecule's integrity depends on the maintenance of the globular structure of this domain. This structure depends on the salt bridges between the positively charged residues of block 4 and its negative counterparts in block 5.

The results of experiments in which fragments of Cry1Aa and Cry1Ac were interchanged and of the interactions between these chimeric proteins and the receptor obtained from *Lymantria dispar* suggest that some residues of Domain III also participate in the interaction with the receptor. It is not known if this is a general function of the domain in the *Bt* toxins.

It is proposed that when δ-endotoxins form pores in the membrane they cause the death of the epithelial cells because they disturb the system that mantains the pH gradient. This inactivation is the result of the alteration in membrane permeability because the target membrane is less permeable to cations. When the motive force that mantains the gradient of 1000 times more protons in the cytoplasm than in the lumen is disturbed, the cytoplasm becomes alkalinized and this disturbs the cellular metabolism causing the destruction of the intestinal tissue. Once this physical barrier is destroyed, the spores obtain access to the haemolymph, where they proliferate because there is plenty of nutrients. The larva dies due to inanition and septicaemia.

PLANTS RESISTANT TO INSECTS

Since the beginning of agrobiotechnology in 1983, *Bt cry* genes have been identified as key elements in order to increase the agricultural productivity, decreasing the use of agrochemicals. Many important crops and different types of plants have been genetically transformed, and in many cases *Bt* genes have been introduced. The main results are reported below.

During the period 1987–1995, 2261 field trials were done with transgenic plants in the USA, in 7095 different locations. From that total, 23.1 % correspond to insect-resistant plants; 27.8 % to herbicide resistance; 26.8 % to improved characteristics; 11.5 % to virus resistance and 2.9 % fungus resistance.

In the USA from 1993 to 1996, 17 transgenic plants were approved for widespread use and production, six of which were resistant of insects (lepidoptera and coleoptera) due to the introduction of the δ-endotoxin *Bt* genes.

TABLE 7.3 : GLOBAL STATUS OF APPLICATIONS FOR THE COMMERCIALIZATION OF TRANSGENCI CROPS WITH *BT* INSECT RESISTANCE

Country	Crop	Year of approval (company)
Argentina	Corn/maize	P (Ciba)
Corn/maize	P (Monsanto)	

Corn/maize	P (Northrup King)	
Australia	Cotton	P (Monsanto)
Canada	Corn/maize	1996 (Mycogen-Ciba)
Potato	1996 (Monsanto)	
European Union	Corn/maize	P (Ciba)
Corn/maize	P (Monsanto)	
Japan	Corn/maize	P (Ciba)
Corn/maize	P (Northrup King)	
Potato	P (Monsanto)	
Mexico	Cotton	1996 (Monsanto)
Corn/maize	P (CIMMYT)	
Potato	1996 (Monsanto)]	
USA	Corn/maize	1995 (Ciba)
Corn/maize	1996 (Northrup King)	
Corn/maize	1996 (Monsanto)	
Cotton	1995 (Monsanto)	
Potato*	1995 (Monsanto)	
Potato	1996 (Monsanto)	

P = pending; * resistance to coleopterans by a *Bt* toxin

Now the methodology for plant transformation of dicotyledons is available and there have been important improvements in the transformation of monocotyledons.

In Mexico the International Center for Improvement of Wheat and Corn has been developing, with financial support from the United Nations, a project to obtain corn varieties with resistance to insects. In Feburary 1996, the first transgenic corn was field tested, containing the *cry*1Ab gene of *Bt*. This field trial was very important because Mexico is the centre of orgin of corn, and the possibility exists of genetic flux between transgenic plants and its ancestor, the teocintle, in this case it was necessary to plan and design a very careful trial and to obtain the approval of the National Agricultural Biosafety Committee of the Secretary of Agriculture. Also it was neccesary to assess the possible implications of introducing transgenic corn in productive areas where teocintle could exchange genetic material with transgenic corn.

Novel Systems Using *Bt*

Recently, the use of *Bt* genes and its proteins has increased, because several research groups have developed novel systems to use and produce them. For example:

- introduction of *Bt* gnes in *Pseudomonas fluorescens* (CellCap technology);
- introduction of *cry* genes into algae to control *Anopheles* mosquitoes (malaria);
- introduction of *Bt* genes into endophytic bacteria in order to control sucking insects (aphids);
- production of *Bt* toxins in baculoviruses as an alternative expression system.

All these efforts are based on the fact that cry proteins are recognized as non-toxic to humans, mammals and other commercially important species. It also means that *Bt* will be used against a wide variety of pests that affect not only agriculture but also other areas such as human health (e.g. in destruction of mosquitoes that transmit malaria).

CONCLUSION

We blieve that the discovery of novel *Bt* activities and strains will continue because there are many insects, mostly in the developing world, that are not effectively controlled with the known cry toxins. Our knowledge about the structure and mode of action of *Bt* toxins will help us to increase the speed of discovery of new cry proteins.

The great challenge of this research field understanding of the mechanism of action of cry toxins in order to develop new strategies to face this problem. The increasing use of transgenic plants makes this challenge bigger.

8

Absorption of Heavy Metal

Metals are an integral part of the Earth's crust, mostly present as insoluble precipitates and minerals. Natural concentrations of metal ions in freshwater systems can be attributed to natural geochemical dissolution, weathering and microbial leaching. Nevertheless, the greater load of metal ions enters the environment as a result of industrial activities such as mining, heavy industry and metal refining.

Although most metal ions are essential components of biological systems, all are potentially toxic. The major physiological functions and mechanisms of toxicity result from their electrical charge, coordinating capacities and the possession of multiple oxidation states. Toxic effects include the blocking of functional groups of biomolecules and functional cellular components, conformational modification, denaturation and inactivation of enzymes and disruption of cell and organellar membrane integrity. However, because micro-organisms possess mechanisms for intracellular accumulation of essential metal ions for growth and metabolism, many other metals having no biological functions can also be accumulated.

Toxic metal ion contamination of the environment is a significant world-wide phenomenon. The recognition of toxic effects from minute concentrations of some metal ions has resulted in regulation laws to reduce their presence in the environment to very low levels as ppb (i.e. $\mu g\ l^{-1}$). The major processes used to treat metal-contaminated waters include the use of ion-exchange resins and the addition of lime to precipitate metal ions. Other, less frequently used, methods include activated carbon adsorption, electrodialysis, and reverse osmosis. However, these traditional technologies are often ineffective or very expensive when used to reduce the presence of heavy metal ions to very low concentrations. Thus, new technologies are required to decrease metal ion concentrations to environmentally acceptable levels at affordable costs.

Bacteria, cyanobacteria, algae, fungi and yeasts are able to remove metal ions from their surrounding environment by means of physicochemical mechanisms – as adsorption – and/or metabolic-dependent activity – as transport. Some physicochemical interactions may be indirectly dependent on metabolism through the synthesis of certain cell constituents or metabolites that may act as metal chelators or create an environment that induces metal precipitation or deposition. In this way, living and non-viable biomass as well as their by-products can accumulate metal ions.

The extent of metal ions removal depends on the physical and chemical characteristics of the water, such as its pH values, temperature, concentration and state of the metal ions, and a number of biological characteristics as well. These include all those parameters which can modify the cell wall composition, such as growth conditions and those parameters which can modify the cell wall composition, such as growth conditions and the age of the culture. Changes of the cell wall may consist in a variation of the number of functional groups or their states, and thus the capacity of metal accumulation is influenced. Furthermore, when using non-viable biomass, mechanical or chemical treatments after cell growth may increase the maximum metal ion amounts removed by adsorption. This enhancement is based on the use of cell constituents from the interior of the cells or break-up of the cell wall and a diffusion into deeper layers of the cell wall.

Native biomass exhibits low mechanical strength, low density, and small particle size. As such, native biomass for metal ions removal must be employed in continuous stirred tank reactors. On the other hand, after metal removal the biomass must be separated from the liquid phase by filtration, sedimentation, or centrifugation. Therefore it is necessary to convert biomass into a form in which it can be used as an ion-exchange resin. Modified biomass must have a particle size similar to that of other commercial resins and possess particle strength, high porosity, hydrophilicity, and resistance to aggressive chemicals. All these properties can be achieved by means of biomass immobilization. Immobilization improves the physical characteristics of biomass for use in reactors, permits continuous operation and protects the microorganisms from microbial degradation.

MICROBIAL REMOVAL OF METAL IONS

Although both living and non-viable cells are able to accumulate metal ions, there may be differences in the mechanisms involved in either case, depending on the extent of their metabolic dependence. Metabolism-independent adsorption of metal ions to cell walls, extracellular polysaccharides, or other materials occurs in living and non-viable cells and is generally rapid. Metabolism-dependent intracellular uptake or transport occurs in living cells and is usually a slower process than adsorption, although greater amounts of metal may be accumulated by this mechanism in some organisms. In growing cultures of microorganisms, metabolism-independent and dependent phases of metal removal can be affected by chelators. Thus, in a given microbial system,several mechanisms of metal removal may operate simultaneously or sequentially.

The choice of living or non-viable biomass for metal ions removal depends on each particular case because both options present advantages and disadvantages. In the case of non-viable biomass, the biological contribution is limited to the choice of conditions of growth giving optimal adsorbent qualities. During metal exposure no nutrient feed is necessary and metal toxicity is unimportant; furthermore, the situation is not complicated by the production of metabolic end products that may complex the metals away from the cells. Nevertheless, biomass becomes quickly saturated and metal desorption is necessary before further use. On the other hand, living cells have advantages in their variety of accumulation mechanisms and relative ease of morphological, physiological and genetical manipulation. Additionally, there is far greater potential for a long-term continuous process without the need for desorption, giving continuous biomass growth and replenishment. However, accumulation of metal ions by living cells is inhibited by low temperatures, metabolic inhibitors and the absence of an energy source, and is influenced by the metabolic state of the cells and the composition of the external medium. In addition, metal ions can be present only at low concentration because of their toxicity towards living cells.

There are six predominant mechanisms by which microorganisms facilitate removal of soluble metal ions from solution: volatilization, extracellular precipitation, extracellular complexing and subsequent accumulation, intracellular accumulation, oxidation–reduction reactions and adsorption to the cell surface.

Volatilization

A number of microorganisms are able to generate volatile organometallic compounds by means of metal methylation reactions. These reactions are thought to be a microbial resistance mechanism to volatilize and thus remove the metal ions from their environment. One well-known example of volatilization is the methylation of selenium, in which the selenate and selenite ions are converted into the volatile selenium compounds dimethyl selenide, dimethyl diselenide and dimethyl selenone. Volatilization is important in metal ions transformation in the environment; however, because of the toxicity of some methylated metals and the difficulties of capturing them, this mechanism has not been considered for developing a commercial process for treating metal-containing waters.

Extracellular Precipitation

Some metabolic products excreted by microorganisms to the environment are capable of immobilizing metals ions. One of the best examples of extracellular precipitation of metal ions is the production of hydrogen sulphide by sulphate-reducing bacteria. These bacterial oxidize organic matter and reduce sulphate to sulphide, which readily reacts with soluble metals to form insoluble metal deposits.

Often the site of metal deposition is unspecified, but with *Citrobacter* sp. metal ions bind as metal phosphate on the cell wall. Metal accumulation is mediated by a surface located acid-type phosphatase. The enzyme cleaves glycerol 1-phosphate or other suitable organic substances and liberates phosphate, which precipitates stoichiometrically with the metal ion to form metal deposits tightly bound at the cell surface.

Extracellular Complexing and Subsequent Accumulation

Some microorganisms synthesize molecules such as siderophores and extracellular polymers that have a high binding efficiency for metal ions. Many microorganims release high-affinity iron-binding molecules called siderophores which are catecol or hydroxamate derivatives. These molecules facilitate uptake of iron into the cell; they are also able to bind other metal ions as gallium, nickel, uranium, thorium and copper.

Many capsular and slime exoploymers of microorganisms act as polyanions under natural conditions. The anionic character of polysaccharides is conferred by the carboxylic acid groups, which are partially ionized at neutral to alkaline pH. Polysaccharides also contain an abundance of hydroxyl groups, which tend to interact with metals. It is quite likely that the metal-binding capacity contributed by the protein component of exopolymeric material involves nitrogen-containing groups such as amino and amide. The two most important types of interactions between metal ions and exopolymers are those that involve salt bridges with carboxyl groups on acidic polymers and those that ivolve weak electrostatic bonds with hydroxyl groups on neutral polymers. Depending on the nature of the organic metal and the immediate chemical environment, the metal can form metallic aggregates. Aggregate formation begins when a metal is precipitated by hydrolysis , changes its oxidation state, or reacts with counterions in solution. Aggregate growth can then proceed by means of crystal nucleation. This kind of metal binding is thought to be the most important mechanism of metal removal in activated sludge.

Intracellular Accumulation

Intracellular accumulation of metal ions is an active process depending on cellular metabolism and usually requires specific transport system, although in some cases intracellular uptake is due to changes in membrane permeability arising from toxic effects. Transport systems encountered in microorganisms are of varying specificity and both essential and non-essential metal ions can be taken up. Most mechanisms of metal transport appear to rely on the electrochemical proton gradient across the cell membrane, which has a chemical component, the pH gradient, and an electrical component, the membrane potential, although K^+ gadient may also be involved. Once inside the cell, metal ions may be located within specific organelles and bound to metal-chelating proteins synthesized to respond to the cytotoxic effects of metal ions.

Oxidation–reduction Reactions

A number of microorganisms are able to actively transform metals by either oxidation or reduction of metal ions, although metal ion reduction can also occur passively after metal binding to reactive sites on and within microbial cells.

Metal oxidation

Chemoautotrophic bacteia obtain energy solely from the oxidation of inorganic substances. One important representative of this group of microorganisms is the bacterium *Thiobacillus ferrooxidans*. Thiobacilli oxidize the sulphides and iron-oxidizing bacteria in turn rapidly oxidize the ferrous ion to ferric iron. *T. ferrooxidans* plays a significant roll in the formation of acidic drainage and can be used to minimize the environmental impact of acid mine drainage. Acidic drainage must be neutralized in order to precipitate soluble metal ions by raising the pH to between 9.5 and 11. The reason is that ferric iron precipitates as a hydroxide at pH 4.3, as opposed to pH 9.5 for rerrous iron, and thus oxidation of the iron substantially reduces the consumption of the neutralizing agent.

Another example is the case of *Leptothrix* spp., microorganisms that derive energy from the oxidation of organic matter, and actively oxidize manganese and iron, depositing the oxides within the sheath that covers the organisms. However, there is no evidence that these microorganisms obtain energy from the oxidation of these metals.

Metal Reduction

Partial metal reductions are thought to be mediated by the microorganism's electron transport system. These partial reductions can transform a metal ion into a form that is less mobile or toxic in the environment. An example is the reduction of Cr(VI) to Cr(III), which is less soluble and toxic, by *Pseudomonas putida*.

In some cases metal ion reduction is complete, rendering the elemental metal. Reduction of mercury ions to Hg(0) is a well-known detoxification mechanism associated with the activity of a mercuric reductase enzyme. Hg(0) is volatile and so lost from the medium.

Metal reduction can also occur passively. Passive metal reduction occurs when metal ions are bound to an intracellular or extracellular component that functions as a reductant. The active reduction of Au(III) to Au(I) and the subsequent passive reduction to Au(0) to form colloidal deposits when Au(I) reacts with the alga *Chlorella vulgaris* has been reported.

Adsorption to Cell Surfaces

Microorganisms can accumulate essential and non-essential metal ions by precipitating or binding the metal ions onto cell walls or cell membranes. The surfaces of microrganisms are composed of macromolecules having and abundance of charged functional groups such as carboxylate, amine, imidazole, sulphydryl, sulphate, hydroxyls. and phosphate. Usually, the net charge of cell surface is negative because of the abundance of carboxylate and phosphate residues, and cationic metals can passively bind to cell surfaces, a process called *biosorption*. However, because of the presence of amine and imidazole groups, which are positively charged when protonated, cell surfaces may also bind negatively charged metal complexes. Thus, micro organisms contain many polyfunctional metal-binding sites for both cationic and anionic metal complexes.

Because adsorption of metal ions onto cell surfaces is a reversible process, the mechanism probably involved is the electrical bonding between the anionic groups of the cell wall and cationic metals, although van der Waals forces. Covalent bonding and redox interactions can also be involved. Electrical attractions usually imply an ion-exchange type of reaction.

Mechanisms of metal ions binding onto cell surfaces fit the adsorption isotherm types L and S described by Giles and Smith (1974). With type L adsorption there is a progressive decrease in the availability of binding sites with increased metal binding and early saturation. In the type S adsorption the presence of bound metal during the early stages of adsorption promotes an increase in the quantities subsequently adsorbed and can be described by a system of multiple equilibria. In this case, whole cells may be necessary for maximal accumulation but, conversely, with accumulation processes relying on type L adsorption mechanisms, extracted polymer may give a higher specific metal accumulation than whole cells.

Cyanobacterial Envelopes

The envelopes of cyanobacteria are similar to those of Gram-negative bacteria. They consist of two membrane bilayers (the outer and cytoplasmic membranes) that sandwich a peptidoglycan layer between them. This peptidoglycan layer is responsible for the rigidity of the cell and for its resistance of osmotic lysis. The peptidoglycan is a meshwork, of linear *N*-acetylmuramyl-(β-1,4)-*N*-acetylglucosamyl strands that are covalently linked together by bonds between short constituent peptide stems. The end result is a skein of glycan fibres that are linked together to form a three-dimensional macromolecule. The outer membrane is firmly bound to the underlying peptidoglycan by a small lipoprotein. This lipoprotein has its lipid, at one end of the lipoprotein, embedded in the outer membrane, while the other end provides a covalent link to the peptidoglycan. The outer membrane has a lipopolysaccharide in the outer side, which is composed of a polysaccharide and a lipid, at one end of the lipoprotein, embedded in the outer membrane, while the other end provides a covalent link to the peptidoglycan. The outer membrane has a lipopolysaccharide in the outer side, which is composed of a polysaccharide and a lipidic moiety called lipid A. Additionally, numerous cyanobacteria possess an envelope outside of the lipopolysaccharide. This is variously called the sheath, glycocalyx, or capsule – or merely gel, mucilage, or slime, depending on its consistency. The sheaths of cyanobacteria are predominantly polysaccharides, but up to 20 % of the weight may be polypeptides and, depending on the species, many types of sugar residues may be involved.

The most probable candidates for metal ion binding are the phosphate groups that are resident within the polar head groups of the lipopolysaccharide and phospholipids of the outer membrane, and the carboxylate groups of the peptidoglycan.

Factors Affecting Metal Ion Absorption

The pH plays a critical role in the binding of metal ions to cell wall functional groups. Most metal ions can be divided into three major categories, depending on how binding to the microorganisms is affected by pH. One group of metal ions, including Hg(II), Au(III) as $AuCl_4$, Ag(I), Pd(II) and Au(I) thiomalate, bind to microorganisms rather independently of pH values between 2 and 7. This behaviour is consistent with the general coordination chemistry of metal ions. These metal ions are all classified as 'soft' and undergo covalent bonding to softer ligand, such as sulphydryl and amine groups. Those bonding interactions are generally minimally affected by ionic interactions and pH.

A second group of metal ions, which binds more strongly to microorganisms as pH increases from 2 to 5, consists of borderline soft and 'hard' metal cations, including Cu(II), Ni(II), Zn(II), Co(II), Pb(II), Cr(III), Cd(II), U(VI), Be(II), and Al(III). A pH dependence of metal cation binding generally occurs when the active metal-binding sites can also bind protons. Thus, metal ions and protons compete for the same binding sites. The incresed binding of these ions at high pH is consistent with electrostatic binding to ligands such as carboxylates and phosphates, which can be negatively charged because of deprotonation at high pH values.

The third group of metal ions binds more strongly to microorganisms at pH 2 than at pH 5. These ions include mainly oxoanions and other anionic metal complexes. The increased binding of these ions at low pH is consistent with electrostatic binding to ligands such as amines or imidazoles, which can be positively charged because of protonation at low pH values.

The temperature dependence of metal ions adsorption process is determined by the enthalpy of the reaction. In general, complex formation between metal ions and carboxylate ligands is characterized by a small positive enthalpy, wherease complex formation between metal ions and amine and sulphydryl ligands exhibits a rather large negative enthalpy. Thus, when a ligand containing an amine and a carboxyl group interacts with metal ions, the heat of reaction is dominated by the negative enthalpy due to the amine. Therefore, the magnitude and sign of the enthalpy for a reaction involving metal ions and a substrate that contains mixed functionalities is a weighed average of the contributions of each ligand. The observed enthalpy in a microorganism–metal ion reaction can therefore be positive, negative, or zero depending on the metal to biomass ratio and the relative extent of binding to different sites.

In general, the presence of two or more metal ions in the medium leads to a decrease in the individual adsorption of each metal ion because of the competition for the same adsorption sites on the cells. However, the observed effect can vary considerably depending on the microorganism, the metal combination, the metal concentration and the order of metal addition due to the existence of different types of adsorption sites on the cell surface, which have preference for binding either hard or soft metal ions. On the other hand, cations usually present in hard waters as Ca(II) and Mg(II) have no effect on the adsorbent potential for other metal ions, which is an important advantage over commercial ion-exchange resins.

The metal ions adsorption process is also affected by the presence of organic and inorganic ligands that can act as metal chelators. These ligands compete with the micro-organisms for the metal ions. The metal–ligand complexes formed away from the cell are not suitable for adsorption. The extent of the interference depends on the metal–ligand complex stability constants.

Desorption of Metals from Microorganisms

The effect of pH and the presence of competing ligands on metal ion adsorption by microorganisms can be used to reverse the process; to desorb and recover the adsorbed metal ions from cells. For metal ions that show a marked pH dependence in binding to cells, desorption can be accomplished by pH adjustment. Metal ions that show little pH dependence in binding to cells can be successfully desorbed by the addition of specific ligands that form exceptionally stable complexes with these ions.

Advantages of Biosorption

Several potential advantages are possible with biosorption processes, including: the use of naturally abundant biomaterials that can be cheaply produced; the ability to treat large volumes of metal-contaminated waters because of the rapid kinetics; the high selectivity in terms of removal and recovery of specific metal ions; the ability to handle multiple metal ions and mixed waste; the high affinity, reducing residual metals to below 1 ppb in many cases; the lower requirement for additional expensive process reagents, which typically cause disposal and space problems; the operation over a wide range of physicochemical conditions including temperature, pH, and the presence of other ions as Ca(II) and Mg(II); the relatively low capital investment and low operational costs; the greatly improved recovery of bound metal ions from the biomass; and the greatly reduced volume of hazardous waste produced.

BIOMASS IMMOBILIZATION

When practical purposes are considered, free biomass presents serious problems compared with its immobilized counterpart, including those involved in biomass manipulation and separation from the processed effluent. Also, practical application of free biomass in biosorption operations often malfunctions because of pressure drops across a fixed-bed column during down-flow operation; this is caused by cell clumping. Excessive hydrostatic pressure is thus required to generate a suitable flow rate. As some of the employed cells, i.e. algal cells are rather fragile attrition of the biomass may occure under high pressure. In summary, free cells are suitable for only a limited number of applications as, for example discontinuous reactors.

At first, these problems seemed to preclude the practical application of biosorption-based processes. However, the referred difficulties can be overcome in the main by using an appropriate immobilization method for the selected biomass. Thus, biomass immobilization provides good handling and operation characteristics to the biomass immobilization provides good handling and operation characteristics to the biomass. It also permits its use in conventional engineering process designs. Listed someother advantages offered by immobilized whole cells in wastewater treatment. However, a number of disadvantages exist that must be also considered. Apart from higher global costs, immobilization often reduces the efficiency of the biosorbent cells by means of the blockage of some functional surface groups. Biomass immobilization decreases mass transfer due to the presence of lower diffusion rates.

Cell immobilization can be defined as the confinement of whole cells in an insoluble phase, which permits the free exchange of solutes from and towards the biomass but at the same time isolates the cells from their surrounding medium.

There are some considerations about the art of immobilization. Immobilization of biomass in a matrix that is too dense can significantly reduce metal loading. Functional groups can be masked by the binding material, as mentioned previously. In the development of economically and technologically viable biosorption products, there must be a balance between a bead with substantial chemical and physical integrity but reduced stability.

If the immobilization is too aggressive, the resulting bead lacks porosity and the contact time between the immobilized biomass and the metal-containing water must be increased to allow for the increased diffusion time required for the metal ion to reach the functional group. This can significantly increase column size.

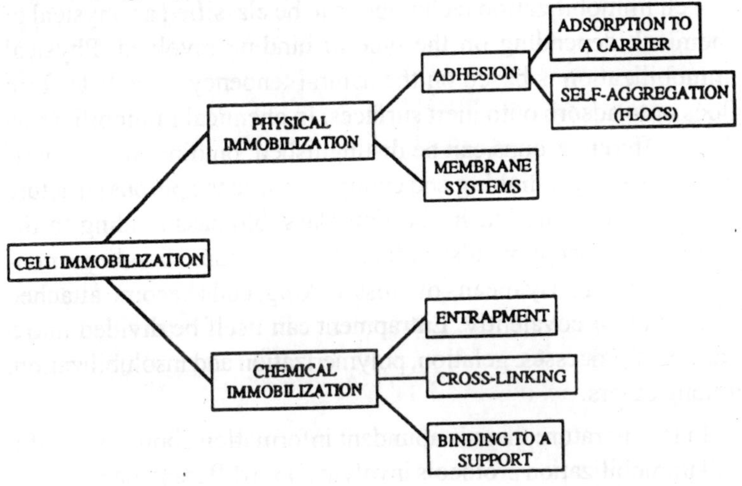

Figure 8.1 : ***The main methods of cell immobilization.***

Metals adsorbed onto the immobilized biomasss can be eluted in the same way that in the case of free biomass, i.e. by means of the appropriate desorbent agent. Often the optimum eluent consists of a moderately acid solution. The most commonly employed acids are hydrochloric, nitric and sulphuric acids. In some cases, pH reduction is not sufficient, or is simply ineffective, for metal removal from the biomass. Certain metals (Au, Ag, Hg) involve different mechanisms in their binding to the surface active sites. So chelating compounds, such as EDTA (ethylenediamine tetraacetic acid), or complexing agents, such as thiourea, turn to be necessary for the recovery of such metals from biomass.

Cell Immobilization Methods

In recent years the immobilization technology of whole cells has developed greatly. A high number of substances has been examined for the immobilization of different types of biomass. For instance, polyacrylamide, calcium alginate, and silica were used as a support for algal immobilization. Other natural materials used as immobilization matrices are agar, agarose, *k*-carrageenan and diatomaceous earth. Polyurethane and polyvinyl foams, polyacrylamide, ceramics, epoxy resin and glass beads are among the synthetic meterials most frequently used as biosorbent supports.

Cell immobilization techniques can be classified as physical or chemical, depending on the type of binding involved. Physical immobilization is based on the natural tendency of cells to form flocs or to adsorb onto inert surfaces. In chemical immobilization three different groups can be distinguished: binding to a supporting surface, cross-linking and entrapment into the porous structure of a polymeric matrix. In the first class, biomass binding to the surface may occur by adsorption, chelation, ionic bonding or covalent bonding. By means of cross-linking, cells become attached to each other covalently. Entrapment can itself be divided into a variety of processes: gelation, polymerization and insolubilization, among others.

In the literature there is abundant information about successful cell immobilization protocols involving both different matrices and biomass from a number of different sources. A broad spectrum of applications is encountered for these immobilized cells. In our laboratory we have improved the methods reported by Ferguson *et al.* (1989) and Klein and Kressorf (1982) to immobilize non-viable biomass of the thermophilic non-nitrogen-fixing cyanobacterium *Phormidium laminosum* by entrapment in microporous beads of polysulphone and epoxy (oxyrane) matrices. Non-immobilized biomass of this filamentous cyanobacterium was previously reported to be an excellent heavy metal biosorbent.

TREATMENT OF METAL-CONTAINING EFFLUENTS

As mentioned above, retention of the biomass in/on a selected matrix permits its application for practical heavy metal biosorption purposes using conventional engineering systems. Most metal-ion removal processes using non-viable immobilized biomass usually involve either a batch or a column configuration reactor. Briefly:

- *Stirred-tank reactors*. In a batch configuration, immobilized biomass is mixed with the metal ions and, following equilibration, is removed by settling or another suitable mechanism.

- *Packed-bed reactors*. Typically, in a column configuration the metal ions are pumped through a column reactor packed with immobilized biomass. The effluent to be treated is pumped from the top of the column. Then the cleaned fluid is pumped out at the bottom of the reactor. Ideal support for this application should have a low cost and a large surface/volume ratio.
- *Fluidized-bed reactors*. In these reactors, influent flow is pumped through the biomass particles from the bottom of the column with a high enough speeds to maintain immobilized cells in a low dense bed. Thus contact area becomes increased, reducing both reactor volume and treatment costs.
- *Dispersed-bed (air-lift) reactors*. Particles of biomass are kept in constant movement by means of air as propellant. As a consequence, particle– solution contact becomes enhanced and so does the global efficiency of the process. Higher costs are the main drawback.

Factors Affecting Metal Biosorption

Apart from the previoulsy mentioned factors affecting metal biosorption when free biomass is considered (pH, surface structure, etc.), a number of other factors can be taken into consideration in the case of immobilized biomass. The most important parameters are contact time, diffusion rate, flow speed, column size and the number of biosorption/elution cycles, among others. These are those normally found in other processes dealing with the same type of reactors.

The kinetics of metal binding to immobilized algae was found to be much slower than the rapid kinetics of metal ion binding by free algal cells. Authors have proposed some ways to improve the binding properties: using a countercurrent feed flow, decreasing the size of the immobilized biomass beads, and increasing the porosity of the beads.

The number of cycles a biosorbent can undergo without losing efficiency appears critical for the overall economics of the system.

Commercially Available Metal Biosorbents

Several products based on immobilized biomass for detoxification of metal loaded wastewaters of effluents have been patented and commercialized. A variety of types of biomass and supports has been employed in the development of such registered products. In some cases, some of the details remain unknown.

BIOCLAIM®

BIOCLAIM® is a metal biosorbent including microorganisms, mainly bacteria of the genus *Bacillus*. Before its immobilization, the biomass is washed with an alkaline solution in order to enhance its metal biosorption, and then rinsed with water to remove excess conditioning agent. Microorganisms so treated are then immobilized in polyethyleneimine and glutaraldehyde or other appropriate binders. Sodium hydroxide, potassium hydroxide, alkaline detergents and other reagents are some of the solutions most commonly empolyed to improve the biosorption capactiy. The caustic treatment is said to provoke ruptures on the cell wall of the microorganisms, exposing new sites for metal binding. Also it solubilizes some cell constituents such as lipids, which do not have extensive metal-binding capacity. However, the most important efffect of the chemical treatment should be to ensure counterions are present in the surface, decreasing protonation. so, beads containing immobilized biomass retain some residual alkalinity, which increases the pH of the metal solution to be treated. As mentioned above, biosorption capactiy increases with pH for some metal ions.

The addition of polyethyleneimine and glutaraldehyde to the prewashed biomass produces a substance with doughlike consistency. This material is readily extrudable through dies of various sizes. This property allows for the production of small (1 mm diameter) beads, which can be rounded using a spherulizing device. The biomass-containing bead, which retains about 60% moisture, does not need to be dried before use. The bead has a specific gravity of 1.3. The BIOCLAIM bead possesses a greater physical integrity than some chelating cation exchange resins. These characteristics permit the utilization of the product in fixed-bed columns (either up-flow or down-flow), fluidized-bed columns and dispersed-bed operations by air forced motion.

The biomass-loaded particles tend to shrink as the metals are accumulated, resulting in a smaller and denser particle. These dense paricles stratify in expanded-bed or dispersed-bed reactors, allowing for selective removal of beads with the higher metal loading. The BIOCLAIM material presents minimal compressibility, thus reducing problems of pressure drop.

Although biosorption materials are relatively non-selective for heavy metal ions, BIOCLAIM beads, and other biosorption products as well, present some metal perference, which is biomass specific. However, when treating mixed-metal wastewaters, the metal with the higher preference did not readily displace metals with lower binding specificities from functional groups.

Metal-loaded BIOCLAIM beads are stripped of metal by means of sulphuric acid, sodium hydroxide for amphoteric metals, or chelating or complexing reagents. The beads are rectivated with a caustic solution before reusing them to extract metals from contaminated solutions.

AlgaSORB

Alga*SORB* is a proprietary family of metal biosorbents that consists of several types of non-living algae and several immobilization matrices. These products are typically employed in fixed-bed reactors for the removal of soluble, heavy metal cations from waste streams. A modification of the product has yielded a new material for the recovery of precious metals.

First-generation AlgaSORB materials were immobilized in silica gel. Like other biomass materials, AlgaSORB removes metals from solution mainly by ion-exchange processes. Once loaded with metal ions the algal products can be stripped of metals to produce a concentrated solution of 10 g l^{-1}. Typically 0.5 M sulphuric acid is used; however, other reagents such as sodium hydroxide can be used when appropriate. Thus, the material can undergo a certain number of biosorption– elution cycles. The number of cycles depends on the conditions employed, and so the economical feasibility of the global process.

Algal biosorbents are particularly useful to remove heavy metals from waters containing high concentrations of Ca(II), Mg(II), Na(I) and K(I), because these ions either do not bind to the biomass surface or are readily displaced by the metal cations of interest. Like other biosorption materials, AlgaSORB is unaffected by the presence of organic contaminants in wastewaters.

BIO-FIX

BIO-FIX is a biosorption process that uses biomass immobilized in polysulfone. The biomass, including sphagnum moss peat, algae, yeast, bacteria, and aquatic flora, is thermally killed and pulverized. The dried and ground biomass is blended into a solution containing a given concentration of polysulfone in dimethylformamide. The mixture is then dropped int water where spherical beads are formed. Nozzle diameter can be varied to control the bead size. Finally, beads are employed in a moistened state to proceed with metal biosorption.

The selection of the biomass used in *BIO-FIX* fabrication is based on the metal accumulation capacity of the biomass and biomass availability. Sphagnum moss peat, the marine alga *Ulva* sp., the cyanobacterium *Spirulina* sp., the yeast *Sacchacromyces cerevisiae*, duckweed *Lemna* sp., xanthan and guar gums, and alginate have been evaluated.

BIO-FIX beads have been demonstrated to be most effective in treating wastewaters with metal concentration ranging from 0.01 to approximately 15 mg l^{-1} soluble metals. Moreover, the low affinity for Ca(II) and Mg(II) is an important commercial aspect of BIO-FIX as well as for other biosorbents. The alkaline earth metals do not readily bind to the biomass functional groups, even when these metals are present in high concentrations. When binding does occur, the alkaline earth ions are easily displaced by heavy metal ones.

Between 75 and 90% of the soluble metal is biosorbed by the BIO-FIX beads in the first 20 min of contact. Because the binding of metals to the functional groups on the adsorption kinetics.

Metal elution from BIO-FIX beads can be accomplished using sulphuric, nitric, or hydrochloric acid. EDTA is not as effective as mineral acids for metal elution. Sodium carbonate was found to be a good conditioning agent for BIO-FIX after metal elution using nitric acid.

Apart from batch or column reactors, BIO-FIX has also been tested for application in a tough system, which used the beads contained in porous bags constructed of fine-mesh woven polypropylene filaments. This system is specifically designed for use at remote sites, such as those with acid mine drainages, where minimal maintenance is required and water flows fluctuate.

Biomass-based Biosorption Processes

Apart from the above-referred products, information from a great number of reported biosorption processes, which are based on a huge diversity of microorganisms, heavy metals and/or conditions, can be found in the literature. What is intended here is not to make and exhaustive review of these studies but only to mention those more remarkable in terms of any particular innovation. Table 8.1 summarized some of the relevant metal biosorption processes using immobilized biomass.

Heavy Metal Biosorption by Phormidium Laminosum

In our laboratory we have recently studied the biosorption of a number of heavy metals by non-viable biomass of the cyanobacterium *P. laminosum* entrapped in microporous beads of polysulfone and epoxy resins and in reactors of different design and operation (Table 8.2). The adsorbed metal can be eluted by washing *in situ* with hydrochloric acid, and after biomass reconditioning with sodium hydroxide the regenerated biosorbent can be used for another metal adsorption cyle. This heavy metal biosorption system can be used for at least ten consecutive adsorption/desorption cycles without apparent decrease of efficiency.

CONCLUSION

Immobilization of biomass of proved metal-loading capability provides a powerful tool for the detoxification of metal-containing effluents. Biosorption of heavy metal ions must be interpreted

TABLE 8.1: IMMOBILIZED BIOMASS FOR METAL BIOSORPTION PROCESSES

Biomass	Immobilization method	Reactor type	Observations	References
Bacillus sp.	Polyethyleneimine and glutaraldehyde	Fixed-bed or fluidized-bed columns	AMT-BIOCLAIM®	Brierley (1991)
Algae	silica gel	Fixed-bed columns	AlgaSORB*	Bedell and Darnall (1990)
Various types	Polysulfone	Stirred-tank, fixed-bed and	BIL-FIX*	Ferguson *et al.* (1989)
Streptomyces albus	Polyacrylamide	Stirred-tank reactor	Uranium removal	Nakajima and Sakaguchi (1986)
Persimmon tannin	formalin	Batch and column	Gold removal	Nakajima and Sakaguchi (1993)
Ascophyllum nodsum	Formaldehyde	Stirred-tank reactor		de Carvalho *et al.* (1994)
Chlorella homosphaera	Alginate	Fixed-bed reacto		Costa adn Leite (1991)
Chlamydomonas reinhardtii and *Selenastrum capricornutum*	Covalent binding to glass (glutaraldehyde)	Minicolumn	Preconcentration for analytical determinations	Elmahadi and Greenway (1994)
Chlorella vulgaris	Polyacrylamide	Column	Selective recovery of metals	Greene *et al.* (1987)
Zoogloea ramigera	Alginate	Bubble column reactor	Three columns in sequence	Kuhn and Pfister (1989)
Rhizopus arrhizus	Reticulated polyester foam	Column (upward flow)		Lewis and Kiff (1988)

Algae	Silica gel	Column	Trace metal preconcentration	Mahan and Holcombe (1992)
Pseudomonas sp.	Polyacrylamide	Column	Uranium uptake	Pons and Fuste (1993)
Bryoria sp.,*Letharia* sp. and	Silica gel	Column	Preconcentration for analytical applications	Ramelow *et al.* (1993)
Saccharomyces cerevisiae	Alginate	Batch and continuous flow systems	Heavy metal and redionuclid e recovery	De Rome and Gadd (1991)
Rhizopus arrhizus	Alginate, polyacrylamide, redionuclide resin and polyvinylformal	Stirred-tand reactor		Tobin *et al.* (1993)
Rhizopus arrhizus	Polyvinylformal	Stirred-tank reactor	Uranium removal	Tsezos and Deutschmann (1990)
Rhizopus arrhizus	Reticulate polyester foam	Stirred-tand and upward—flow column		Zhou adn Kiff (1991)
Scenedesus quadricaude	Cross-linking (ethyl acry late and ethylene glycol dimethacrylate)	Column		Harris and Ramelow (1990)
Saccharomyces cerevisiae	Sand filter	Column (upward-flow)	Huang *et al.* (1990)	
Fungal biomass	Paper filters and polyester fleece	Column	Wales and Sagar (1990)	

TABLE 8.2 : METAL ADSORPTION BY NON-VIABLE BIOMASS OF *PHORMIDIUM LAMINOSUM* IMMOBILIZED IN POLYSULFONE AND EPOXY RESIN IN PACKED-BED AND FLUIDIZED-BED REACTORS

Reactor	Flow rate (ml min^{-1})	Metal adsorbed (mg) in beads of							
		Polysulfone				Eposy resin			
		Cu(II)	Fe(II)	Ni(II)	Zn(II)	Cu(II)	Fe(II)	Ni(II)	Zn(II)
Packed-bed	0.25	5.5	4.62	3.51	6.32	10.23	8.46	7.59	10.39
Packed-bed	0.50	4.59	4.88	4.03	4.94	8.42	7.15	6.46	8.21
Packed-bed	1.00	3.86	3.93	3.54	3.89	6.16	5.56	5.01	6.36
Packed-bed	2.00	3.51	3.24	3.26	3.21	4.69	4.33	3.38	5.04
Fluidized-bed	–	4.28	3.64	4.06	3.75	8.66	6.95	6.01	8.59

Biomass was immobilized by entrapment in polysulfone or in epoxy resin by a modification of the methods reported by Ferguson *et al*. (1989) and Klein and Kressdorf (1982), respectively. Uniform-sized beads of diameter 2,71 (±0.02) mm and 2.12 (±0.02)mm, were obtained. The beads (250 mg biomass dry weight) were packed in column reactors operated in continuous mode or in an air-lift reactor operated in batch. 500 ml of 50 ppm metal were either pumped downwards through the column reactors at the indicated flow-rates or incubated for 36 h in the air-lift reactor

TABLE 8.3 : METAL ADSORPTION/DESORPTION CYCLES BY *PHORMIDIUM LAMINOSUM* BIOMASS IMMOBILIZED IN POLYSULFONE BEADS

Cycle	Adsorbed metal (mg)				Desorbed metal (mg)			
	Cu(II)	Fe(II)	Ni(II)	Zn(II)	Cu(II)	Fe(II)	Ni(II)	Zn(II)
1	3.97	3.34	3.59	4.30	3.76	2.46	2.52	3.52
2	5.07	4.42	4.38	5.39	4.55	2.38	3.79	4.70
3	4.85	4.06	4.32	5.50	4.92	2.37	4.12	4.80
4	4.85	4.30	4.49	5.39	4.80	2.12	3.78	4.65
5	4.70	4.13	4.29	4.73	4.64	1.93	3.67	4.74
6	4.89	4.32	4.28	5.04	4.71	2.10	3.40	4.34
7	4.61	4.19	4.21	5.00	4.66	1.70	3.78	4.35
8	4.84	4.09	4.37	5.06	4.89	1.84	3.45	4.73
9	5.23	4.23	4.19	4.82	4.56	1.70	3.58	4.51
10	4.85	4.19	4.27	4.91	4.70	1.88	3.72	4.47

500 ml of 50 ppm metal were pumped downwards through the column reactors containing 250 mg (dry weight) of immobilized biomass. Bound metal was stripped from biomass with 50 ml of 0.1 M HCl. After each desorption step, the biomass was reconditioned with 30 ml of 0.1 M NaOH. A flow rate of 0.5 ml min^{-1} was used

TABLE 8.4 METAL ADSORPTION/DESORPTION CYCLES BY *PHORMIDIUM LAMINOSUM* BIOMASS IMMOBILIZED IN POLYSULFONE BEADS

Cycle	Adsorbed metal (mg)				Desorbed metal (mg)			
	Cu(II)	Fe(II)	Ni(II)	Zn(II)	Cu(II)	Fe(II)	Ni(II)	Zn(II)
1	4.58	5.32	4.17	5.49	3.93	3.45	3.90	4.67
2	3.86	2.07	1.95	2.06	3.77	2.48	2.30	2.23
3	3.74	1.21	2.49	2.20	4.07	2.06	2.05	2.62
4	3.73	1.52	2.79	1.89	3.94	1.73	2.16	2.38
5	3.74	1.27	2.81	2.62	3.82	1.53	2.16	2.38
6	3.18	1.60	2.46	2.56	3.28	1.64	2.23	2.30
7	3.80	1.46	2.67	2.22	3.67	1.42	2.05	2.36
8	3.73	1.13	2.62	2.47	3.56	1.43	2.07	2.26
9	3.95	1.21	1.93	2.36	3.56	1.23	1.95	2.50
10	3.51	1.25	2.27	2.58	3.64	1.10	1.98	2.41

500 ml of 50 ppm metal were pumped downwards through the column reactors containing 250 mg (dry weight) of immobilized biomass at 0.5 ml min^{-1} flow rate. Bound metal was stripped from biomass with 170 ml of 0.1 M HCl pumped at the same flow rate. After each desorption step, the biomass was reconditioned with 30 ml of 0.1 M NaOH at 2 ml min^{-1}

in terms of a complementary and emergent technology to conventional wastewater treatments. The broad and diverse number of practical applications of such biosorbents found in the literature gives and idea of the growing importance of this novel technique. However, a certain period of field testing seems to be necessary before this tool can be broadly implemented.

9

Biological Removal of Heavy Metals

Although nitrates occur naturally, they have become important pollutants as a consequence of human activities. In fact, due to the use of fertilizers in excess and the increasing accumulation rate of wastes from human and animal populations, many sources of drinking water contain high levels of nitrate ion.

The major sources of nitrogenous wastes can conveniently be considered in two groups, chemical and biological. the main chemical sources are fertilizer production, explosives manufacture, coal conversion and other industries such as the manufacture of sodium carbonate, petroleum refining, refrigeration plants using ammonia in scouring and cleaning operations, certain synthetic fibre, plants, etc. Biological sources of nitrogenous pollutants are sewage, animal husbandry, food and fermentation industries, natural sources (decaying vegetation, animal and bird droppings).

The best-known effect of fertilizer run-off and the other forms of nitrate pollution is eutrophication of water, which in extreme cases can result in water becoming clogged with algae, killing most other plants and animals. Regarding nitrogen pollution, most concern stems from the possible health hazards that have been attributed to nitrate, either directly as a causative factor of methaemoglobinaemia, or indirectly as the source of carcinogenic nitrosamines. However, the possible damage to the ozone layer of the Earth's atmosphere by gaseous oxides of nitrogen, released from nitrate ion in the soil, is also considered an important aspect of air pollution.

Although to date it does not seem to present any problem for human health, phosphorus removal from municipal and industrial wastewaters is reuired to protect water bodies from eutrophication, especially in lakes and enclosed bays with stagnant water. Municipal, agricultural and forest wastes are each responsible for about 30% of the total known sources of phophates, with detergents contributing about 60% of the municipal phosphate effluent.

MICROALGAE AND CYANOBACTERIA

Photosynthetic microorganisms can be grouped into two categories: photosynthetic bacteria and microalgae. Photosynthetic bacteria perform anoxygenic photosynthesis and possess bacteriochlorophyll, a light-absorbing pigment chemically distinct from the chlorophyll *a* present in all other photosynthetic organisms (algae and higher plants).

Cyanobacteria (also known as blue–green algae or cyanophyceae) constitute the largest, most diverse and most widely distributed group of photosynthetic prokaryotes. they hold an intermediate position between photosynthetic bacteria and eukaryotic algae, but they do not possess either bacteriochlorophyll or chlorophyll *b*. In contrast to photosynthetic bacteria, cyanobacterial perform oxygenic photosynthesis similar to that of higher plants.

From a biotechnological point of view, the term 'microalgae' (which has no taxonomic meaning) includes those microorganisms that possess chlorophyll *a* and other photosynthetic pigments capable of performing oxygenic photosynthesis. Therefore, two different cellular types are combine within such a term: cyanobacteria and eukaryotic microalgae.

The large-scale cultivation of microalgae was probably first considered seriously in Germany during World War II, with the aim of production lipids by means of growing *Chlorella pyrenoidosa* and *Nitzschia palea*. This initial research was taken up by a group of scientists at the Carnegie Institution of Washington in order to use the green alga *Chlorella* for large-scale production of food. Since then, many research teams in several countries have been successfuly developing techniques for the cultivation of microalgae on a large scale, mostly freshwater ones such as *Chlorella, Scenedesmus, Coelastrum, Dunaliella* and *Spirulina*. some cyanobacteria, such as *Spirulina*, are among the most commonly utilized organisms in microalgal biotechnology.

Although there are unmistakable indications that photosynthesis is the principal mode of energy metabolism in these organisms, in the natural environment they regularly experience dark diurnal periods and some can even survive long periods in complete darkness. Consequently, they possess a respiratory metabolism capable of providing maintenanc energy in the dark.

Cyanobacterial have a world-wide distribution and most species are cosmopolitan. They may occur in extreme habitats such as hot springs, in desert rocks and below the Antarctic ice and are known to be amongst the earliest colonizers of arid land. They were probably largely responsible for our oxygen-enriched atmosphere and the roots of the cyanobacterial lineage extended far into the geological past. Owing to the longevity of the lineage and its influence on the Earth's early environment, cyanobacteria must rank as among the most successful and ecologically significant of the myriad forms of life to have appeared over the history of biological evolution.

BIOLOGICAL WASTEWATER TREATMENT

Biological removal of nitrogen appears a valid option and offers some advantages over chemical and physicochemical treatments: it can operate under relatively wide pH variations (at least in the case of using microalgae), it does not create secondary pollution problems, and the biomass can be exploited commercially.

The low capital and operational costs of biological processes for removal of phosphorus have also been attracting attention in recent years. The biological process offers the advantages of not requiring addition of chemicals and of reducing the volumes of sludge produced. Recently, instead of the chemical precipitation methods, the biological phosphorus removal capability of the anaerobic/oxic process has been substantiated experimentally.

The application of high-rate algal ponds in wastewater treatment appears to have the greatest potential of all biotechnologies based on microalgae, if exploited fully as a multipurpose system. Although the most widely utilized biological method for wastewater treatment has focused mainly on removal by bacteria, over the last 20–25 years increasing attention has been paid to the possibility of using microalgae. In fact, wastewaters from domestic, industrial or agricultural sources are a favourable microalgal culture medium. The idea of using microalgae for this purpose, as initially proposed and experimentally tested by Caldwell (1946) and, some 10 years later, by Oswald and Gotaas (1957), has gained momentum and a fair number of papers have dealt with such systems.

At present, one of the promising fields using microalgae in wastewater treatment is the utilization of cyanobacterial. Through the appropriate use of the nutrient uptake capabilities of these prokaryotes, nutrients in the secondary effluent of wastewater treatment plants can be removed and used in the growth of algal biomass, which can then become a source of biomass. Such light-driven biotechnology may provide an effective means of removing inorganic nitrogen (ammonium, nitrate, nitrite) and phosphorus (phosphate) in primary and secondary urban wastewater treatment processes where the ions are implicated in the eutrophication of the receiving bodies of water.

In countries with a high number of hours of sunlight there has been considerable development of combined algal and bacterial systems for wastewater treatments. However, treatment of wastewaters with microalgae is possible under harsh conditions by using greenhouse technology. At present, algal treatment systems operate with high efficiency in various parts of the world.

The cost of microalgal cultures can be at least partially overcome by selling the biomass or extracting high added-value products. However, large-scale solar biotechnology is still confronted with the expense of harvesting microalgal biomass.

CYANOBACTERIA

In general, cyanobacteria can use nitrate or ammonium as the sole nitrogen source for growth, and may have the additional ability to fix atmospheric nitrogen (N_2). In the utilization of any form of inorganic nitrogen, first, the nitrogenous compound in the outer medium should enter the cell and, second, since ammonium is the only nitrogen form directly incorporated into amino acids, dinitrogen and nitrate must be reduced to ammonium. Finally, ammonium is incorporated as an α-amino group into carbon skeletons. Nitrate is probably the most common source of nitrogen for utilization by microalgae and cyanobacterial in the environment and only the ionized form is encountered at physiological or ecological pH values. In addition, nitrate assimilation is affected by a number of environmental and nutritional factors such as light, temperature, pH and carbon source availability. Under autotrophic conditions, the utilization of any form of inorganic nitrogen by cyanobacteria is strictly dependent on the availability of light and CO_2. Light is needed to synthesize the ATP and reductant required for the assimilation processes involved, whereas the CO_2 requirement results from a set of interactions between carbon and nitrogen metabolism.

Orthophophate is believed to be the form of phosphorus commonly taken up bty cyanobacteria. Phosphate flow into the cell is closely linked to photosynthetic ATP formation and thus depends strongly on the overall phosphorylation potential of the cell. The transport of phosphorus through the cell membrane appears to be the rate-limiting step in its incorporation.

The intracellular accumulation of phosphate is energy dependent, being higher in the light than in the dark, and depends strictly on the pH of cytoplasm and not on the energy conversion at the thylakiod membrane, which is responsible for the energy supplies.

Finally, the major phosphate reserve of cyanobacteria is polyphosphate, which accumulates as discrete granules in the cytoplasm of the cell when phosphate is in excess. Polyphosphate is mobilized during periods of nutrient shortage, representing a valuable pool of activated phosphorus, which can be used in a variety of metabolic processes.

IMMOBILIZATION TECHNIQUES

One of the major problems in use of microalgae for the biological treatment of wastewaters is their recovery from the treated effluent. In order to solve the problem of harvesting the biomass, many systems have been proposed or tried, but only two appear feasible: chemical flocculation, especially with chitosan and bioflocculation. Among the most recent ways of by passing this problem are immobilization techniques applied to algal cells. In fact, in the case of photosynthetic cells, over the last 20–25 years considerable progress has been achieved in the immobilization of photobiological organisms and organelles such as cyanobacterial, photosynthetic bacteria, algae and chloroplasts.

Cell Immobilization

Rosevear (1984) defined immobilization as a technique which confines a catalytically active enzyme or cell within a reactor system and prevents its entry into the mobile phase, which carries the substrate or product. According to Fukui and Tanaka (1982), two different types of immobilized cells can be distinguished:

- Immobilized, treated cells: cells utilized in a dead state being subjected to an appropriate treatment before or after the immobilization. However, the desired enzymes are in an active and stable form.
- Immobilized, living cells: either resting or growing in gel matrices.

Although the advantages (and drawbacks) of using immobilized rather than free-living cells depend on the intrinsic properties of the cells and the purpose of their use, some general observations can be made. Immobilized cells may offer certain specific advantages over batch or continuous culture fermentation where free cells are used, such as accelerated reaction rates due to increased cell density per unit volume, increased cell metabolism and cell wall permeability, no wash-out of cells, high operational stability and better control of catalytic processes, separation and reuse of catalyst, reduction in cost due to easier separation of cells and excreted product.

There are three major techniques to be used for cell immobilization: entrapment, adsorption and coupling. The immobilization techniques which most resemble the circumstances in which cells find themselves in nature, are their entrapment within gels and adsorption to surfaces. In fact, many organisms normally exist adsorbed to surfaces while entrapped in a gel or slime of their own making.

Entrapment

In the case of entrapment, the cell is merely restricted in its movements and confined to a small volume of liquid within a defined microenvironment. The matrix must be constructed *de novo* around the cells rather than adsorbing the cells onto a preformed material. One of the major problems found when entrapping cells in a gel is the diffusional resistance of the gel to the substrate and products. Another important problem that arises when using entrapped cells lies in their ability to divide and eventually break the support. The gelation of alginic acid by polyvalent metal ions is one of the most important methods of cell immobilization. The technique involves the drop-wise addition of cells suspended in sodium alginate onto a solution of calcium chloride to form a very stable gel where the cells are intrapped.

A number of polyurethane prepolymers are commercially available and have been used successfully to entrap cells. Urethane is formed by the reaction between an isocyanate and a hydroxyl group. Condensation of the urethanes with other isocyanates, alcohols or amines produces a cross-linked polymer. By varying the temperature, prepolymer structure, condensation reagent, among others, polyurethane foams of different porosity, strength and translucency can be produced. Addition of cells to prepolymers before condensation can result in a uniform distribution of the biomass in the foam.

Adsorption

Adsorption of cells onto a solid surface is probably the mildest of the cell immobilization techniques. Many organisms are capable of naturally adhering to surfaces; in fact, whenever a suspension of cells is brought into contact with a surface there will almost always be a certain amount of adsorption to the solid.

Adsorption thus describes an immobilization method in which the biocatalyst is attached to a surface by non-covalent interactions. The types of adsorption can range from non-specific binding dependent on weak surface force through the strong interaction of ionic bonds to the highly specific, multipoint binding of natural biological affinities. Cells are capable of producing extracellular polysaccharides once adsorbed on a surface and this may further strengthen the hold of a biocatalyst to the matrix.

Polyurethane and polyvinyl foams have been used as supports for immobilized cells.

Coupling

Covalent coupling is perhaps the most popular technique when immobilizing enzymes, but few systems using this technique on cells have been reported. This is because it is a generally harsh technique, which results in loss of viability of cells. In addition, where viable cells can be immobilized by covalent coupling, any cell division is likely to result in cell leakage from the support, as the daughter cells will be probably less strongly attached.

Coupling is based on covalent bond formation between activated support and cells. Another possiblity is to covalently cross-link the cells to one another, providing greater stability to the aggregates that can be achieved by flocculation. The most widely used coupling agent is probably glutaraldehyde although carbodiimine, isocyanate and aminosilane have also been used.

Microalgae have been used successfully for the removal of nitrogen and phosphorus pollutants from water. Chevalier and de la Noue (1985a,b) have shown that cells of *Scenedesmus* sp. immobilized on *k*-carrageenan beads are as efficient as free cells in taking up ammonium and orthophosphate from secondary urban effluents. *Phormidium* sp. cells were attached to chitosan flakes and used for removing nitrogen (ammonium,nitrate, nitrite) and orthophosphate from urban secondary effluents found that *Chlorella emersonii* entrapped in calcium alginate beads was able to remove phosphorus from secondary treated effluent with acceptable efficiencies.

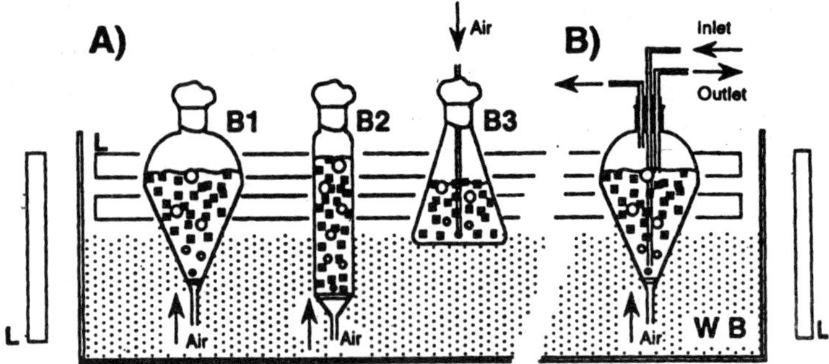

Figure 9.1 : ***Schematic representation of the air-agitated bioreactors used : B1, funnel shape; B2, Erlenmeyer flask; the liquid phase volumes were 250, 125, and 250 ml, respectively. (A) Batch fluidized-bed; (B) Continuous-flow air agitated. L, Fluorescent lamp; WB, water bath. The liquid phase onsisted of ultrapure water supplemented with 50 mg l^{-1} $NaNO_3$ or 10 mg l^{-1} Na_2HPO_4. Preboiled, washed and dried 5-mm foam cubes were placed in a cell suspensiuon and immobilization by adsorption was carried out. Once the foam cubes were fully colonized, they were introduced into the bioreactors. Effluent samples were collected at timed intervals and the ion concentration determined. The reactor was continously illuminated with cool white fluorescent lamps at a light intensity of 100 µmol photon m^{-2} s^{-1}. Algal leakage from the react9r was identified by the presence of free cells in the effluent and was measured by absorbance at 678 nm. The B1 bioreactor was also tested in a continous-flow mode (B) by connecting it to a supply tank containing ultrapure water supplemented witht he same ion concentrations. Teh water was supplied at the top of the bioreactor using an on-line variable-speed peristaltic pump at different flow rates. Uptake efficiency (UE) in the continous-flow bioreactor was defined as: UE = [(I – E)/I] × 100, where I and E were the influent and effluent concentrations of the ions, respectively. An efficiency value of 100% was obtained when no ion appeared int eh effluent (i.e. when E = 0).***

Garbisu *et al.* (1991, 1992, 1993, 1994) reported on the utilization of the filamentous thermophilic cyanobacterium *Phormidium laminosum* immobilized in polymer foams for the removal of nitrate, nitrite and phosphate from water. In view of several studies, it appears that the filamentous cyanobacteria *Phormidium* spp. may be among the most promising microalgal genera for the processes of biological tertiary treatment used in the depollution of water. In fact, *P. laminosm* shows some characteristics that,in an immobilized state, make it a suitable choice for the removal of inorganic nitrogen and phosphorus from water: it is a non-N_2-fixing cyanobacterium that can utilize nitrate, nitrite, or ammonium ions as the sole nitrogen source for growth; it can stand wide variations of pH (6–11) and temperature (15–50°C) and can grow at relatively low light intensities; its filamentous nature is an advantage when immobilizing the cells be entrapment or by adsorption into matrices; its hydrophobicity and tendency to attach to surfaces gives it a good potential in processes involving immobilized cells; and its metabolism can be readily altered by nutritional stress.

Consequently, we carried out some studies to study the feasibility of using polymerimmobilized *P. laminosum* for the removal of inorganic nitrogen and phosphorus from water. In order to do so, batch and continuous-flow air-agitated and packed-bed bioreactors were used to study the removal of nitrate, nitrite and phosphate from ultrapure water by polyvinyl-immobilized *P. laminosum* cells as a preliminary approach to the applicability of these systems for the removal of ions from polluted potential drinking water.

The results showed that *P. laminosum* cells immobilized in polyvinyl foam and packed in a column reactor took up nitrate continuously for at least 3 months. During this time, the cells remained active and nitrate uptake efficiencies in excess of 90% were achieved.

We studied the effects of light intensity and CO_2 concentration on this packed-bed bioreactor performance with respect to the nitrate uptake efficiency of the system. From these studies, it was concluded that not only CO_2 concentration but also light intensity affects nitrate uptake, although the effect of the latter must be probably associated with the CO_2 fixation rate. In fact, the limiting factor in nitrate utilization appears to be the availability of CO_2 fixation products, the light dependence of the process being a reflection of the light dependence exhibited by CO_2 fixation. Thus, at limiting light intensity, a moderate competition between nitrate utilization and CO_2 fixation occurs as both processes require reducing power and/or ATP; resulting in a decrease in the rate of CO_2 fixation in the presence of nitrate. However, at high light intensities and saturating CO_2 concentrations, no competition for assimilatroy power between nitrate assimilation and CO_2 fixation occurs because the photosynthetic apparatus is able to generate enough assimilatory power for both simultaneous processes.

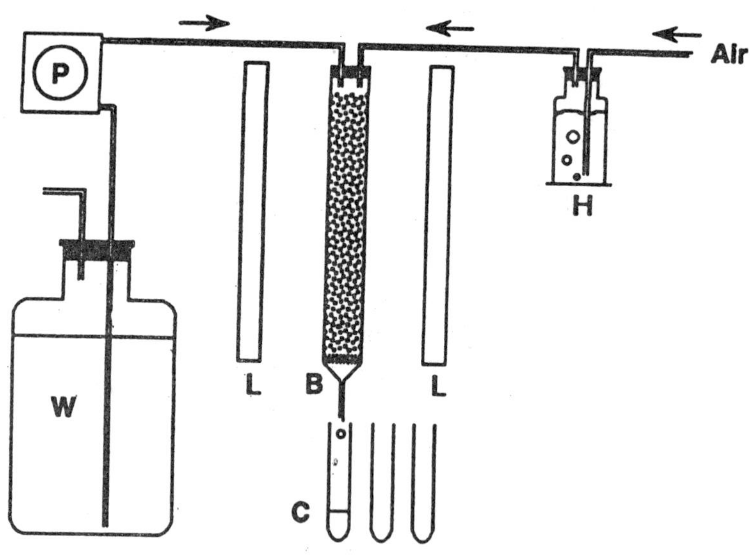

Figure 9.2 : ***Schematic diagram of the continous-flow packed-bed bioreactor. L, Fluorescent lamp; B, bioreactor; H, humidifier flask; C, collector; W, water (feeding tank); P, pump. A glass column (2.7 cm inside diameter, 50 cm long) was used. Previously colonized foam cubes were packed int he bnioreactor and the column was connected to a feed tank containing ultrapure water supplemented with*** **50** ***mg*** l^{-1} **$NaNO_3$** ***or*** **10** ***mg*** l^{-1} **Na_2HPO_4*****. The medium was supplied at the tip of the column using a peristaltic pump at different flow rates. Column effluent samples were collected at timed intervalsw and the ion concentration determined. The column was continously illuminated with cool white fluorescent lamps at a light intensity of*** **100** ***µmol photon*** m^{-2} s^{-1}***. Uptake efficiency and cell leakage wre identified as in the continous-flow air-agitated bioreactor.***

Continuous-flow air-agitated bioreactors showed nitrogen removal efficiencies of up to 90% for residence times of 14 h in short-term experiments. Although nitrogen-starved cells showed higher inorganic nitrogen-uptake rates than nitrogen-sufficient ones, the photosynthetic activities of the cells decreased progressively with the time of nitrogen starvation. Nitrogen deficiency can be easily induced by incubating the cells for a given time in a nitrogen-free medium. Although the incubation of cells in a nitrogen-free medium up to about 70 h resulted in higher uptake rates, longer incubations led to lower uptake rates, probably due to the excessive degeneration of the cell structures. These nitrogen-starved cells produced high amounts of exopolysaccharides, which appear to assist the immobilization process. The fact that after approximately 60 h of nitrogen starvation none of the photosynthetic activities (photosystem I, II or I + II) could be detected might be due either to the photosystems having been rendered inactive or to the high amounts of exopolysaccharides produced by the nitrogen-starved cells which could act as an additional diffusion barrier to the entrance of the electron mediators used to measure these activities into the cells and to gas exchange.

If nitrogen-starved cells were to be used practically for the removal of nitrogen and phosphorus from water, several bioreactors would need to be set up in parallel, alternating cycles of nitrogen starvation and supply, to keep the system operating continuously with starved cells.

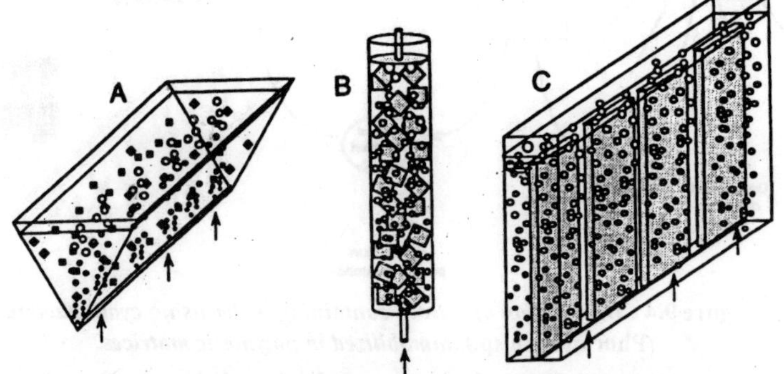

Figure 9.3 : ***Schematic diagram of the coninous-flow air-agitated photobioreactors of different shapes used for the removal of nitrate from water. (A) Triangular reactor; (B) column reactor; (C) rectangular reactor. Air was supplied to the cultures through the bottom of the bioreactors (arrows).***

Inorganic nitrogen uptake by nitrogen-starved cells occurred in both light and dark under aerobic conditions. In anaerobiosis light was required for the uptake, confirming that the necessary energy might perhaps be derived from the respiratory electron transport chain under aerobiosis. The uptake of nitrate and nitrite in the dark and in the absence of an added carbon source by nitrogen-starved cells could be due to these cells accumulating high carbohydrate reserves during the starvation period, which can apparently substitute for photosynthetic CO_2 fixation products in stimulating ion uptake.

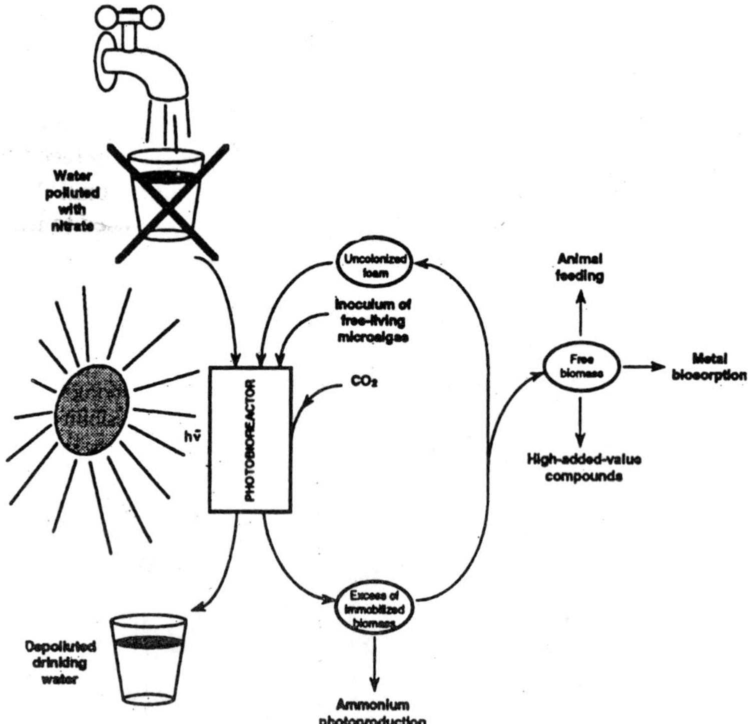

Figure 9.4 : *Depollution of nitrate-containing water using cyanobacteria (*Phormidium sp.*) immobilized in polymeric matrices.*

Ammonium inhibited nitrate uptake but did not affect the uptake of nitrite. Nitrate uptake appears to be regulated by a negative feedback control exerted by certain compounds produced during nitrate assimilation and ammonium metabolism. The effect of ammonium upon nitrate uptake may be seen as an exaggeration of this feedback control continuously modulating the uptake rate. The fact that nitrite uptake was not inhibited by ammonium appears to support the idea that two different transport systems (permeases) operate in the uptake of nitrate and nitrite ions.

Blanco *et al.* (1993) also studied the nitrate-removal capacity from water of immobilized microalgae and found results similar to those reported by Garbisu *et al.* In their studies, as well as *P. laminosum* cells they utilized other microalgae such as *Phormidium uncinatum, Scenedesmus obliquus* and *Phormidium* sp. in different air-agitated photobioreactors of various shapes. In all those works, and using residence times of 3–5 h and light intensities of 100 μmol photon m^{-2} s^{-1}, they found nitrate uptake efficiencies of up to 90% of the concentration of this ion supplied in the influents to the bioreactors (50 mg l^{-1} nitrate).

By contrast, *P. laminosum* cells immobilized in polymer foams did not show high phosphate uptake efficiencies. Nitrogen-starved cells were also examined in relation to their phosphate uptake characteristics and it was shown that starvation led to lower uptake rates. The addition of nitrate to nitrogen-starved cells markedly increased phosphate uptake. Phosphate uptake was inhibited in the dark. Although our data showed that nitrogen-sufficient *P. laminosum* had singnificant phosphate uptake only when light was available, this should not be an insuperable obstacle when a bioreactor is designed to deal with the practical problem of removing phosphate from polluted water. In fact, given the likelihood that light will not penetrate very far into the columns of immobilized algae, various ways have been used to solve this light penetration problem (which always aries when using photobioreactors). Approaches studied are: matrices with internal reflections. organisms which enhance light penertation (e.g. tubular photobioreactors).

When considering the possibility of using these immobilized systems for the removal of ions from polluted water, the fact that no significant leakage of cells was observed in the effluent during reactor operation is of vital importance. Lack of cell leakage over comes one of the major problems on the utilization of microalgae for the biological treatment of water, i.e. the recovery of microalgae from the treated effluent. Once the supports are fully colonized and the cells begin to leak out from the foams into the effluent, the bioreactor can be stopped and most of the immobilized cells easily removed from the foam matrices by squeezing and the foams subsequently reused without reinoculation. If cells cannot be recycled due to their age or metabolic state, they can be utilized for different purposes, e.g. feed (fish), extraction of high-added value compounds (phycobilin pigments), or removal of heavy metals by means of chemically treated biomass.

From all these studies, it was concluded that *P. laminosum* immobilized on polymer foams is of potential value for biological removal of inorganic nitrogen and phophorus from water in continuous-flow bioreactor systems.

CONCLUSION

Biological removal of nitrogen and phosphorus from water appears to offer some advantages over chemical and physicochemical treatments and, therefore, has been attracting attention in recent years. In this context, although the most utilized biological methods for wastewater treatment have focused mainly on removal by bacterial, during the last 20–25 years numerous investigations have been carried out on the possibility of using microalgae. Wastewaters from domestic, industrial or agricultural sources are a favourable microalgal culture medium. Through the appropriate use of the nutrient uptake capabilites of these organisms, nutrients (such as inorganic nitrogen and phosphorus) in the secondary effluents of wastewater treatment plants can be removed and used in the growth of algal biomass.

So for, one of the major problems in the use of microalgae for the biological treatment of wastewaters is their recovery from the treated effluent. Among the most recent ways of bypassing the problem of harvesting the biomass are immobilization techniques. Consequently, various studies have been carried out to study the feasibility of using immobilized microalgae (mainly cyanobacteria immobilized in different matrices) in batch and continuous-flow bioreactors for the removal of inorganic nitrogen and phosphorus from water. From all these studies, it was concluded that some polymer immobilized cyanobacteria are of potential value for biological removal of inorganic nitrogen and phosphorus from water in continuous-flow bioreactor systems.

10

Biological Treatment of Polluted Soil

Bioremediation is defined as the use of biological treatment systems to destroy, or reduce the concentrations of, hazardous wastes from contaminated sites. Such systems have potentially numerous applications., including clean-up of ground water, soils, lagoons, sludge and process waste streams. Bioremediation has been used on very-large-scale applications such as the shoreline-clean-up efforts in Alaska resulting from the Exxon Valdez oil spill in 1989. At the same time bioremediation is still considered an innovative technology which has been used in only a limited number of cases, most of them in USA. Several advantages of bioremediation systems have been identified. It:

- can be done on site
- keeps site disruption to a minimum
- eliminates transportation costs and liabilities
- eliminates long-term liability
- uses biological systems, often less expensive
- can be coupled with other treatment techniques into a treatment train.

Some authors consider that bioremediation is still unproven and that there are some disadvantages and/or limitations when it is applied to specific problems, such as:

- some chemical compounds are not biodegradable

- extensive monitoring is required
- each site has specific requirements
- potential production of toxic unknown subproducts
- strong scientific support is needed.

This chapter describes in detail the main aspects and components of bioremediation system.

CURRENT MARKET FOR BIOREMEDIATION

Bioremediation comprises today only a small fraction of the very large market for hazardous waste treatment, but it is one of the fastest growing sectors in environmental management. In the USA particularly the bioremediation industry is still limited, but progresses toward commercialization at a faster rate that in other countries. The key reasons for this situation are:

- Most bioremediation research and development have been carried out in the USA.
- The scope and enforcement of USA environment law exceeds that of other nations.
- Public acceptance of bioremediation in the USA is high due to the publicity generated by the use of bioremediation for clean-up of large oil spills.

There are several estimations about the world market for bioremediation, indicating that the most important sector of the bioremediation market is petroleum-contaminated soils and groundwater, resulting from leaking underground storage tanks (USTs). In Europe, within the next 10 years soil clean-up costs alone are estimated to exceed US$30 billions; if 5 % of this soil is cleaned using biotreatments then 1.5 billion dollars will be spent by bioremediation methods. In the USA there are around 750 000 existing UST facilities, over 50% of which are used for storage of petroleum hydrocarbons, which are biodegradable. Conservative estimates indicate that one out of three of these tanks is leaking and if 10% of these sites undergo biological clean-up the total revenue would be US$4 billion. Other future markets for biological treatment include process waste pretreatment, industrial lagoons, municipal landfill leachates and general chemical spills.

The potential world-wide revenues for bioremediation systems are US$ 11.5 billion over the next decade.

In practice the situation is different. In a review of the Superfund Program Remedial Actions it was found that nearly half of the remedial treatment for source control (primarily soils) involves technologies that were not available when the Superfund Program was authorized in 1986. The study, which was performed in 1995, evaluated more than 300 innovative technologies projects for soil and groundwater, the most frequent being solidification/solubilization (29%), followed by off-site incineration (15%) and on-site incineration (11 %). Bioremediation (*ex situ* and *in situ*) accounts for only 10 %, but among innovative technologies was second only to soil vapour extraction (19 %).

BIOREMEDIATION SYSTEMS

In order to carry on bioremediation, several biological systems and techniques have been developed in the last few years (Table 10.1). The diversity of bioremediation technologies is an indication of the complexity that one encounters when trying to clean soils contaminated with some hazardous organic matter.

TABLE 10.1 : BIOREMEDIATION TREATMENT TECHNOLOGIES

Bioaugmentation	Addition of bacterial cultures to a contaminated medium; frequently used in bioreactors and *ex situ* systems
Biofilters	Use of microbial stripping columns to treat air emissions
Biostimulation	Stimulation of indigenous microbial populations in soils and/or ground water; may be done *in situ* or *ex situ*
Bioreactors	Biodegradation in a container or reactor; may be used to treat liquids or slurries
Bioventin	Method of treating contaminated soils by drawing oxygen through the soil to stimulate microbial growth and activity

Composting	Aerobic, thermophilic treatment process in which contaminated material is mixed with a bulking agent; can be done using static piles, aerated piles, or continuously fed reactors
Land farming	Solid-phase treatment system for contaminated souls; may be done *in situ* or in a constructedd soil treatment cell

The selection of a bioremediation system is not a simple task because there are many choices and it is solved on a case-by-case basis. We will not describe in detail any single bioremediation system; instead we will discuss their main general components and charcteristics.

The successful implementation of bioremediation techniques will involve a multidisciplinary approach requiring knowledge from individuals with expertise in chemistry, microbiology, geology, environmental engineering, chemical engineering, soil science, etc. In order to solve an environmental contamination problem using bioremediation, it is necessary to study three important aspects: microbial systems, type of contaminant and geological and chemical conditions at the contaminated site.

TABLE 10.2 : CLASSIFICATION OF MICRORGANISMS

Group	Carbon source	Energy source
Photoautotrophs	Carbon dioxide	Light
Photoheterotrophs	Organic carbon	Light
Chemoautotrophs	Carbon dioxide	Inorganic chemical (e.g. nitrate)
Chemoheterotrophs	Organic carbon	Light

Microbial Systems

The key players in bioremediation are microscopic organisms that live virtually everywhere. Microorganisms are the catalyst generator, and the enzymes are the catalysts. These enzymes are involved in degradative (catabolic) reactions to provide energy and material for synthesis of additional microbial cells. Microbial transformation of organic contaminants usually occurs because the organisms can use the contaminat for their own growth as a source of carbon and energy. The mechanisms by which microorganisms obtain energy are the basis for their major classification.

The chemoheterotrophic microorganisms are mainly responsible for the degradation of organic contaminants in the environment. Chemoheterotrophs are a large group of organisms containing numerous genera and species.

Microorganisms gain energy by catalyzing energy-producing chemical reactions that involve breaking chemical bonds and transferring electrons away from the contaminant through oxidation–reduction reactions. The organic contaminant is oxidized (i.e. it loses electrons); correspondingly, the chemical that gains the electrons is reduced. The contaminant is the electron donor, while the electron recipient is the electron acceptor. The energy releasee in these reactions is conserved in the form of the high-energy-phosphate bond of ATP for use in fuelling biosynthetic reactions. In general, the biochemical process can be divided in two groups: fermentation and respiration. These can be distinguished by the nature of the redox reaction on the basis of the terminal electron acceptor.

Fermentation

In fermentation, the microorganisms use the same organic compound as electron donor and electron acceptor. This process does not lead to complete oxidation of the entire original substrate to carbon dioxide. Fermentative metabolism is characterized by production of a mixture of end products with different levels of oxidation, such as organic acids, alochols, hydrogen and carbon dioxide.

Aerobic Respiration

This is the process of destroying organic compounds with the aid of molecular oxygen as the electron acceptor. In these conditions, microorganisms use oxygen to oxidize part of the carbon in the contaminant to carbon dioxide; the rest of the carbon is used to produce new cell mass.

Anaerobic Respiration

Anaerobic respiration occurs only in the absence of molecular oxygen. In these conditions, nitrate (NO_3^-) sulphate (SO_4^{2-}), iron (Fe^{3+}), manganese (Mn^{4+}) and even carbon doixide serve as electron acceptor to degrade contaminants. In addition to new cell mass, the by-products of anaerobic respiration may include initrogen gas, hydrogen sulphide, reduced forms of metals, and methane, depending on the electron acceptor.

Reductive Dehalogenation

Reductive dehalogenation is a variation of the microbial metabolism, which is potentially important in the detoxification of halogenated organic contaminants. Microorganisms catalyse a reaction in which a halogen atom (such as chlorine) of the contaminant molecule is replaced by a hydrogen atom. The reaction adds two electrons to the contaminant, thus reducing it. In most cases, reductive dehalogenation generates no energy but is an incidental reaction that may benefit the cell by eliminating a toxic compound. Most dehalogenated products tend to be less toxic and more susceptible to further microbial decay than the parent compounds.

Cometabolism

In somecases, microorganisms can transform contaminants even though the conversion reaction yields no benefit to the cell. The term for such non-beneficial biotransformations is secondary utilization or cometabolism. Cometabolism is defined as degradation of a compound only in the presence of other organic material that serves as the primary energy source. Cometabolism is the predominant mechanism for the transformation of many substrates, including some polynuclear aromatic hydrocarbons, halogenated aliphatic and aromatic hydrocarbons and pesticides. The organisms that carry out cometabolic reactions include species of *Pseudomonas, Acinethobacter, Norcadia, Bacillus, Mycococcus, Methylosinus*, and *Arthrobacter* among bacteria and *Penicillium* and *Rhizoctonia* among the fungi.

Bioremediation Oganisms

Microorganisms carry out biodegradation in many different environments. Of particular relevance for pollutants are sewage-treatment systems soils, underground sites for disposal of chemicals wastes, groundwaters, etc. Natural communities of microorganisms in these various habitants have an amazing physiological versatility; they are able to metabolize and often mineralize an enormous number of organic molecules. Certain communities of bacteria and fungi metabolize a multitude of synthetic chemicals; the number of molecules that can be degraded is not known but thousands are known to be destroyed as a result of microbial activity in one environment or another. Hazardous compounds result in the selection of a mixed microbial population with improved abilities to tolerate and extract energy from these contaminants. Among the bioremediation microorganisms that frequently are identified as active members of microbial consortiums are: *Acinethobacter, Actinobacter, Alcaligenes, Arthrobacter, Bacillus, Berijerinckia, Flavobacterium, Methylosinus, Mycobacterium, Mycococcus, Nitrosomonas, Nocardia, Penicillium, Phanerochaete, Pseudomonas, Rhizoctonia, Serratia, Trameters and Xanthobacter.*

Microbial Consortiums

In nature there is a diversity of types of microorganisms and energy sources. This diversity makes it possible to break down a large number of different organic chemicals. This capacity is used to fill the needs of the microbial population when degrading complex organic compounds or when a site is contaminated with a mixture of organic materials. Microorganisms individually cannot mineralize most hazardous compounds. Complete mineralization results in a sequential degradation by a consortium of microorgansims and involves synergism and cometabolism actions. As initial transformations often depend on the efficiency of reactions that remove intermediates, the complete degradation of a contaminant is related to the nature of the microbial consortium.

Adaptation

Often, microorganisms do not degrade contaminants upon initial exposure but may develop the capability to degrade the contaminant after prolonged exposure. Adaptation is important because it is a process to ensure the existence of microorganisms that can use xinobiotics which were recently created and introduced in the environment. Adaptation may be defined as the length of time between the entry of a chemical into an environment and evidence of its detectable loss. Adaptation occurs not only within single communities but also among distinct microbial communities that may evolve a cooperative relationship in the destruction of toxic compounds. It is generally desirable to enhance the microbial have already acclimatized to the waste material. Only if the environment is sterile, or the present microbial population does not degrade the contaminant, are exogenous microorganisms considered.

Biological Process Requirements

Microorganisms carry out biodegradation in many different environments. To deal with the assortment of systems of biodegradation, it is necessary to satisfy serveral conditions for degradation to take place in an environment. A summary of biological process requirements are: (a) the presence of organisms with the capability to degrade the target compound or compounds; (b) the chemical substrate must be accessible to the organism and capable of being used as energy and carbon source; (c) the presence of an inducer to cause synthesis of specific enzymes for the target compounds; (d) the presence of an appropriate electron acceptor– donor system; (e) environmental conditions that are adequate for enzymatically catalyzed reactions (moisture and pH); (f) nutrients necessary to support the microbial growth and enzyme production (nitrogen and phosphorus are essential); (g) a temperature range that supports microbial activityand catalyzed reactions; (h) absence of toxic substances; (i) presence of microorganisms to degrade metabolic products; (j) presence of microorganisms to prevent transit build up of toxic intermediates; (k) environmental coditions that minimize competitive organisms with those conducting the desired reactions. It is important to adjust these biological requirements as well as the environmental conditions in order to have a successful bioremediation.

Microbial Nutrition

Microorganisms require food, water, and a suitable environment in which to live, grow and multiply. Food for microbes must provide a source of carbon for synthesis of essential biochemical and cellular components. The carbon source may consist of organic carbon of many varieties (including hydrocarbons), or inorganic carbonate or carbon dioxide. Bacteria used in bioremediation are the hydrocarbon-degrading heterotrophs and those which can use inorganic carbon and salts.

Microorganisms involved in the degradation of xenobiotics in natural systems are dependent on fixed forms of nitrogen (nitrate, nitrite, ammonia or organic nitrogen) to meet their requirements. These forms of nitrogen are frequently the limiting factor for microbial populations in soil.

Phosphorus is essential to microbial cells for the synthesis of ATP, nucleic acids and cell membranes. Because of the low water solubility of phosphates, phosphorus is limiting to bacterial growth in soil environments. In general it is suitable to introduce in culture media in optimal carbon–nitrogen and carbon–phosphorus ratios.

Type of Contaminant

The variety of materials and processes used in industrial activities causes different types of contamination in soils. As a rule, the contaminants are not found individually but in simple or complex mixtures. The mixtures may be associated with the release, storage or transport of many chemicals in surface or groundwater, waste-treatment systems, soils or sediments. The number of chemicals found to date is enormous, and the types of mixtures are similarly countless. Sites are frequently contaminated with complex mixtures of organic compounds such as creosote, petroleum or combinations of industrial solvents. Concentrations of individual contaminants may vary significantly within the site, with some areas having extremely low concentrations and others having concentrations over a millionfold higher. Simple hydrocarbons degrade faster than complex hydrocarbons.

Organic Contaminants

Most of the organic chemicals that are found in the soil system are a result of human activities. These include agricultural amendments (such as fertilizers, sewage, sludge, pesticides), atmospheric deposition (such as automobiles, power plants, municipal incinerators, paint manufacturing industries, etc.), oil-fields brines (hydrocarbons from pertoleum activities, spills, etc.), inert-fill materials (including foundry sand waste, municipal incinerator ash, power plant ash and manufacturing and processing aggregates), xylene, ethyl benzene and chlorinated organic chemicals are often detected in organic solvents used for commercial and household products.

Inorganic Contaminants

- Metals, such as copper, nickel, zinc and cadmium, are often detected in contaminated soil in varying concentrations. The toxic effects are observed at higher concentrations, resulting in the inhibition of metabolic activities of the microorganisms. Microorganisms can not destroy metals but they can alter the microbial reactivity and mobility.
- Cyanide contamination results mainly from disposal of plating bath wastes and plating shop waste. In acidic environment cyanides release toxic hydrogen cyanide. The cyanide ion is a non-specific enzyme inhibitor but exerts its powerful toxic effects by inhibiting the enzyme cytochrome oxidase, and thus preventing the uptake of oxygen by living cells.
- Sulphate contamination mainly results from various human activities with sulphur and its compounds, ranging from acid rain to the disposal of sulphur wastes.

Physicochemical Properties

Determination of the chemical properties of the contaminant is necessary to assess its fate and transport potential, as well as to evaluate its potential for bioremediation. These properties include density, adsorption coefficient for soils, solubility in water, solubility in various solvents, volatilization from soil, volatilization from water, etc.

- *Electron acceptors.* All biological reactions that yield energy are redox reactions. Thus adequate electron acceptors are important to control for bioremediation. Type of electron acceptor establishes the metabolism and therefore the specific degradation reactions. From thermodynamics considerations, the electron acceptors are preferred in a definite order: oxygen, nitrate, sulphate, carbon dioxide, organic chemical (Flectcher, 1994).

- *pH.* For most microbiological activities pH will need to be kept in the range 6–8. The pH affects the microorganisms' ability to conduct cellular functions, cell membrane transport and equilibrium of catalyzed reactions.
- *Temperature* has a marked influence on the rate of bioremediation because bioremediation temperature affects the rate of degradation. There are specific ranges in which the microbial metabolic pathways and enzymes can operate. Each microorganism has a minimum temperature below which growth no longer occurs. Some organisms have the ability to adapt to temperature changes. In general, microorganisms commonly found effective in bioremediation perform well in the temperature range of 10–40°C.
- *Moisture* is an essential parameter of bioremediation. Moisture content of soils affects the availability of contaminants, the transfer of gases, the effective toxicity level of contaminants, the movement and growth stage of microorganisms and species distribution. During bioremediation, if the water content is too high, it will be difficult for transfer oxygen into the soil, and can be a factor that limits growth efficiency. In general, biodegradation of contaminants in soil systems is optimal at a soil moisture content of 10–20 %. This translates to a value of about 40–75 % of field capacity.

Bioavailability of Contaminant

The poor availability or the total lack of bioavailability of a compound may be a major determinant of persistence. Summarize the reasons for the persistence of an organic chemical in soil: (a) incompatibility of environmental conditions with microbial growth; (b) insuffiicient nutrients for degraders to grow; (c) easily degradable carbon is available and causes catabolic repression of degrading enzymes; (d) chemical compound unavailable for the microorgaqnisms; (e) biophysical factors are unfavourable for the degrading enzymes.

In general, adsorption of organics to soil particles reduces the availability of these materials to microorganisms and hence reduces their biodegradability. For instance, microbial degradation of water-insoluble compounds such as pertoleum hydrocarbons require the microbial production of an emulsifying agent to increase the solubility and hence bioavailability of these compounds. Sanseverino reported that surfactant enhanced solubility and biodegradation of polynuclear aromatic hydrocar-bons in coke waste.

We present a gas chromatogram obtained in our laboratory when a naturally selected consortium isolated from a petroleum-contaminated soil was incubated with refined oil in liquid medium. After 3 weeks, under controlled conditions, many of the components of the oil were degraded, showing that bacteria uses hydrocarbons according to their complexity and structure.

Site Characterization

Information about site characteristics should be examined for evaluating the viability of bioremediation technology. The factors, which should be studied, include the chemical characteristics of contaminants as well as the hydrogeological characteristics of the site. A summary of these factors is presented in Table 10.3.

Geological considerations should include stratigraphic effects such as horizontal extent of the aquifer and heterogeneity of the soil, Hydrogeological data include porosity, permeability, and groundwater velocity, direction and recharge/ discharge. In addition, hydraulic connection between aquifers, potential recharge/ discharge areas, and water table fluctuations must be considered. It is important to initially identify the contaminants present and their concentrations, because the microbial systems capable of biotransformation and rates are compound specific.

CONCLUSING REMARKS

In bioremediation, xenobiotics comprise naturally occurring and synthetic compounds that, by nature, only a small part of which

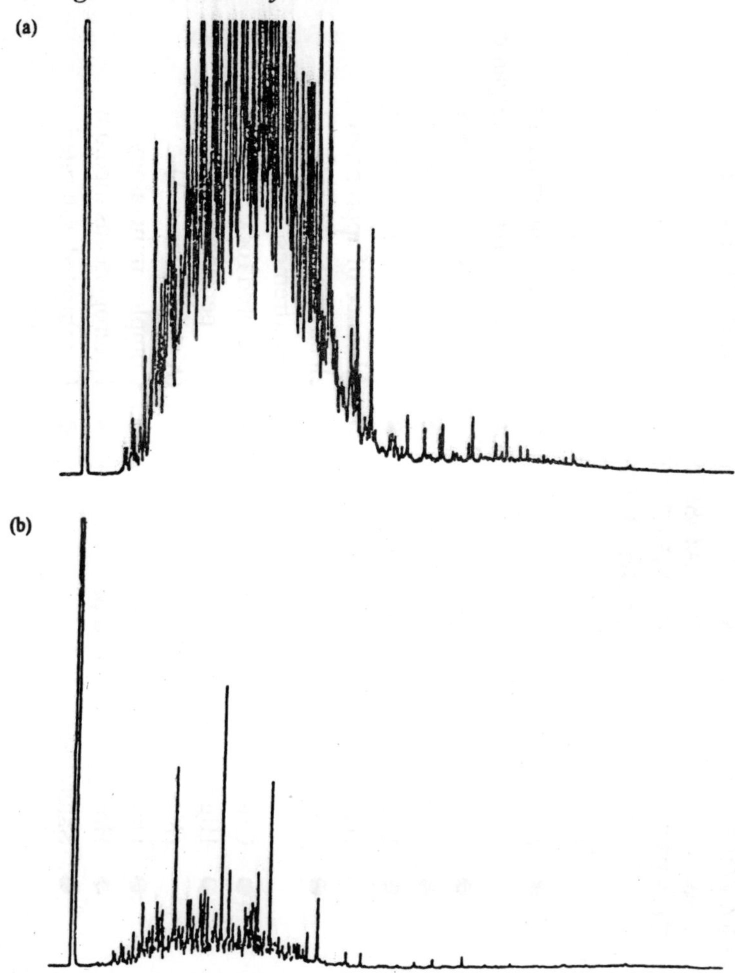

Figure 10.1 : ***Gas chromatogram of refined oil in liquid medium incubated with a mixture of soil microorganisms. The solution was supplemented with nitrogen and phosphorus (a) at 0 h; (b) after 3 weeks.***

can be biodegraded. Under the proper environmental conditions, most xenobiotics are biodegraded by microorganisms already present in the environment. The persistence of xenobiotic compounds is a consequence of the difficulties of the microbiota to reach them. In addition to the bioavailability, a number of environmental factor can affect both the rate and decrease of toxicity

TABLE 10.3 : FAVOURABLE AND UNFAVOURABLE CHEMICAL AND HYDROGEOLOGICAL SITE CONDITIONS FOR IMPLEMENTATION OF *IN-SITU* BIOREMEDIATION

	Favourable factors	*Unfavourable factors*
Chemical characteristics	• Small number of organic contaninants • Non-toxic concentrations • Diverse microbial populations • Suitable electron acceptor condition • pH 6–8	• Numerous contaminants • Complex mixture of inorganic and organic compounds • Toxic concentrations • Sparse microbial activity • Absence of appropriate electron acceptors • pH extremes
Hydrogeological characteristics	• Granular porous media • High permeability ($K > 10^{-4}$ cm s^{-1}) • Uniform mineralogy • Homogeneous medium • Saturated medium	• Fractured rock • Low permeability ($K < 10^{-4}$ cm s^{-1}) • Complex mineralogy • Heterogeneous medium • Unsaturated–saturated conditions

when a contaminant is biodegraded. The ability to manipulate all of these parameters will improve current bioremediation efforts by increasing their effectivity while decreasing cost of the treatment. In order to overcome these limitations, researchers must try to develop more competitive bioremediation technologies. The areas of research include the development of methods for increasing the bioavailability of persistent materials; development of better qualitative and quantitative methods of sampling and analysis of the contamination; improved and standardized techiques for assessement of biodegradation methods and microbial systems; standarized methods to detemine levels of toxicity and monitoring toxic compounds during and at the end of the treatment.

Bioremediation technologies offer a cost-efective, permanent solution to clean-up of soils cotaminated with xenobiotics compounds. It is a new and exciting field and its multidisciplinary nature is a challenge for those interested in the remediation of contaminated soils.

11

Bio-Treatment of Water

The treatment of aqueous effluents by their application to wetlands for natural purification is a historical practice but over the past 15 years considerable interest has developed in the concept of using constructed wetlands for the treatment of point sources of pollution. This interest has arisen because of the need to identify low-cost environmentally friendly techniques for the treatment of dirty waters. Such waters arise from many sources: some contain only a single pollutant, but most contain a wide variety. Materials can be pollutants by adding excess nutrients such

TABLE 11.1: POLLUTING STRENGTH OF DIRTY WATERS

Strength	BOD (mg l^{-1})	Source
Weak	< 45	Effluent from secondary sewage treatment
Medium	45–300	Effluent from primary sewage treatment
Strong	300–3000	Farm dirty water, industrial effluents
Very strong	> 3000	Farm manure slurries, silage liquor, industrial effluents

Dirty water strength can be measured by the biochemical oxygen demand (BOD)

TABLE 11.2 : MAJOR POLLUTANTS

Carbon	Readily biodegradable compounds measured as BOD Slowly biodegradable plus readily biodegrdable compounds measured as chemical oxygen demand (COD) Nitrogen in several forms Measured as total-N, organic-N, NH_4-N, NO_3 and NO_2-N
Phosphorus in several forms	Measured as total phosphate and orthophosphate
Suspended solids	
Heavy metals	Such as iron, manganese, lead, zinc
Pathogens	Measured in colony forming units (cfu) g^{-1} dry weight or wet weight

as carbon substrate, nitrogen and phosphorus, or by causing toxicity by the addition of heavy metals and pathogenic organisms. The work of closely followed by, showed that significant removal of pollutants is possible when contaminated water is passed through beds of reeds planted in soil or gravel. These dirty waters can have a wide range of strengths depending upon their source. The major pollutants present in such contaminated waters range from readily biodegradable compounds to heavy metals.

Basis of Treatment

The aquatic plants mainly employed in constructed wetlands are the common reed, *Phragmites australis*, the common reedmace, *Typha latifolia*, and the common club-rush, *Schoenoplectus lacustris*. These pants have a root system of rhizomes of thick hollow air passages from which the fine hair roots hang down. The vertical aerial shoots develop upwards from the rhizome. In hot climates the floating water hyacinth *Eichhornia crassipes* has been widely used for water purification.

Oxygen from the leaves passes down through the stems and

rhizomes and is exuded from the fine roots so that a thin oxygenated aqueous film surrounds the hairs; as a result this 'root zone' or rhizosphere supports a very large population of aerobic microorganisms, far larger than in comparable soil. In the spaces between the root hairs anaerobic microorganisms predominate, with their slower growing lifestyles.

TABLE 11.3 REMOVAL STEPS IN REED BED TREATMENT SYSTEMS

Water constituent	Removal steps
Suspended solids	Sedimentation, filtration, adsorption
BOD	Sedimentation and filtration
	Degradation to CO_2, H_2O and NH_3 by microorganisms attached to plant and sediment surfaces
Nitrogen	Main removal by nitrification–denitrification
	Ammonia oxidized to nitrite/nitrate by nitrifying bacteria in aerobic zones
	Nitrates converted to N_2 gas by denitrifying bacteria in anoxic zones
	A little removal by plant uptake and ammonia volatilization
Phosphorus	Adsorption, complexation, precipitation reactions within the bed matrix, particularly with aluminium, iron, calcium and clay minerals. Very little plant uptake
Heavy metals	Precipitation reactions after pH changes, sedimentation and adsorption onto biomass films on plant stems
Pathogens	Sedimentation and filtration. Competition and natural die-off. Excretion of antibiotics from roots of plants and from composting of plant litter on bed surface

As a result, water passing through a thickly developed 'root zone' of such aquatic plants encounters alternate aerobic and anaerobic microbial populations which convert the carbonaceous, and to a lesser extent the nitrogenous and phosphorous, contaminants in the water to less polluting meterials. In addition the root growth of the plants causes minor displacement of the media in which the plants are growing and opens up passages for water flow. Above ground the multiplicity of upright stems, often up to 500 per m^2, provides a veritable jungle for water flowing over the bed surface. Microorganisms can form biofilms around the lower stems, which can trap suspended particles in the water by adsorption. Table 11.4 indicates the numbers of microorganisms present in the rhizosphere and on the supporting gravel media.

TABLE 11.4 COMARISON OF MICROORGANISMS FOUND IN GRAVEL MATRICES AND RHIZOSPHERES OF SUB-SURFACE FLOW WETLANDS

Substrate	Bacteria	Actincomycetes	Fungi
Unplanted gravel[a]	0.6×10^6	3.1×10^4	1.0×10^3
Typha gravel[a]	1.6×10^6	1.4×10^5	4.0×10^3
Typha rhizosphere[b]	3.5×10^9	2.5×10^6	2.8×10^4
Phragmites gravel[a]	1.2×10^7	2.8×10^5	1.1×10^5
Phragmites rhizosphere[b]	0.6×10^9	1.3×10^6	1.6×10^6

[a]Expressed as cfu g^{-1} of dry weight
[b]Expressed as cfu g^{-1} of wet weight

It shows that microbial populations in the rhizospheres of *Phragmites* and *Typha* are two to three orders of magnitude greater than those in unplanted gravels, indicating the importance of the vegetation, and that the population is dominated by the bacteria. However, Hatano *et al.* (1993) found that although the bacteria are highest in numbers they have relatively low enzymatic activities within the organic substrates tested. As such they would be expected to carry out most of the degradation of the simpler organic materials in wastewaters. In contrast the actinomycetes and fungi, though fewer in number, have a wide range of hydrolyis activities. Many isolates of these organisms showed significant amylase, protease, chitinase, xylanase and cellulase activity. They would be expected to degrade many of the larger molecules in the wastewaters, which contribute to the chemical oxygen demand (COD) level, in addition to the plant litter falling onto the bed surface.

Horizontal subsurface flow of dirty water through the 'root zone' is used for tertiary treatment of sewage and other weak wastewaters. Downflow through a multi-layered aggregate bed is employed for treating stronger wastes. Above ground flow through the palnt stems is used for metal removal and pH change, particularly for acid mine discharges. Finally, water transpiration and 'root zone' oxidation is used for stabilizing and drying sludges and slurries.

In parctice the common reed, *Phragmites australis,* has been the aquatic plant mostly used in European applications. The common reedmace, *Typha latifolia*, has been extensively used in the USA and is starting to be employed in schemes in the UK using above ground flow for metals removal. These two species have proved well able to tolerate weak, medium and strong wastewaters. So far there have been no major attempts to use reed beds for very strong effluents, as these are likely to put the plants under stress. Moreover, the simple technology of reed beds cannot reasonably be expected to treat this type of material without greatly improved aeration ability.

HORIZONTAL FLOW BEDS

Horizontal flow beds consist of either a soil or gravel matrix into which the reeds are planted. The matrix is kept flooded, with the water surface less than 5 cm below the top of the bed. A basis of design, developed from experience of treating sewage, has been established.

For the tertiary treatment of sewage and for other weak wastewaters the bed length is short, 5–10 . For the secondary treatment of sewage having a BOD of about 300 mg l^{-1} the bed length can be about 70 m; the bed may need to be subdivided into terraces. The residence time in these horizontal flow beds is several days.

At present the design equations available are based on the work of Kickuth. The area of the bed A_h is given by the equation

$$A_h = \frac{Q_d(lnC_o - lnC_t)}{k_{BOD}}$$

where A_h = surface area of bed, m^2; Q_d = daily average wastewater flow rate, m^3/day; C_o = average BOD of inlet wastewater, mg l^{-1}; C_t = required average BOD of outler water, mg l^{-1}; k_{BOD} = rate constant, m/day.

The value of k_{BOD} depends on the biodegradability of the wastewater and on the bed matrix. Values of this constant are given in Cooper *et al.* (1996). For sewage recent UK experience suggests that an area of 3–5 m^2 per person equivalent is appropriate. The equation is of a similar form to design equations for conventional trickling or percolating 'filters'; it can be derived by considering the bed to be a plug flow reactor and assuming a model for the microbial kinetics.

The cross-sectional area of the bed, A_c, can be calculated from the application of Darcy's law for flow of a fluid through a granular bed:

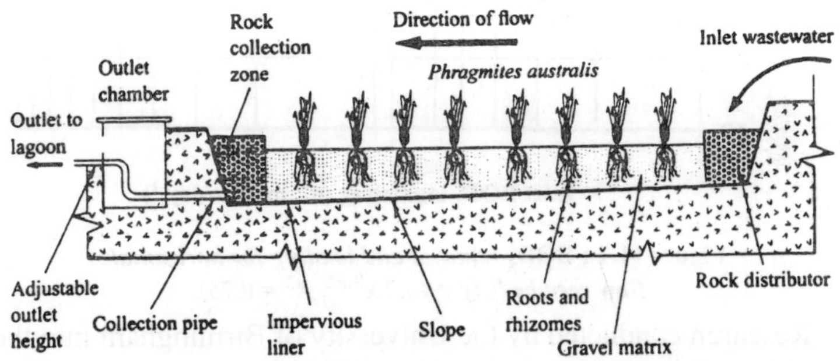

Figure 11.1 : ***Horizontal flow reed bed treatment system showing Phragmites australis planted in a matrix of limestone chippings, the inlet wastewater distribution and outlet purified water collection zones of roceks, and the water level control device. The bed is constructed in a shallow excavation with an impervious liner to prevent seepage of polluted water into the subsoil.***

$$A_c = \frac{Q_s}{k_f dH / ds}$$

where A_c = cross-sectional area m^2; Q_s = average flow rate of wastewater m^3s^{-1}; k_f = hydraulic conductivity of the bed matrix, $m^3m^{-2}s^{-1}$ is used. Kickuth also recommended that the horizontal velocity of the wastewater should be kipt below 10^{-4} ms^{-1} in order to prevent erosion and disturbance of the mosaic of aerobic, anoxic and anaerobic zones in the bed.

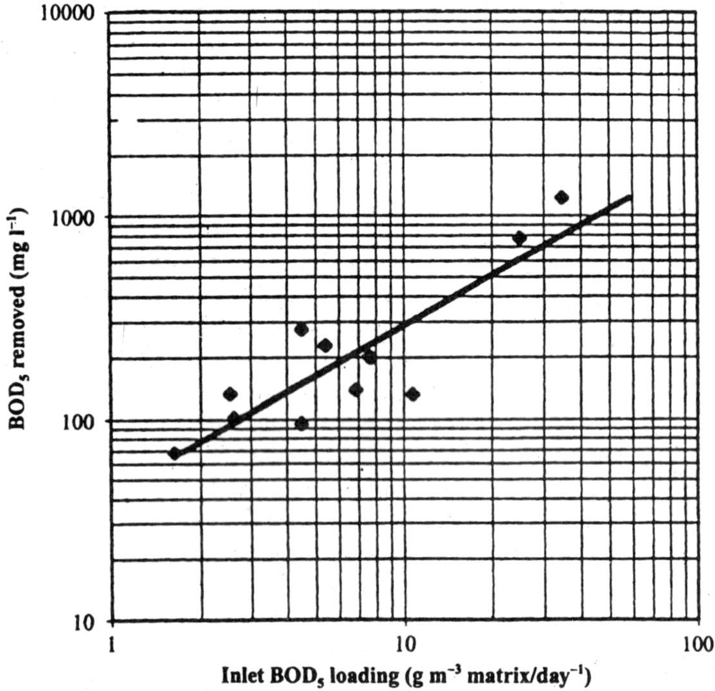

Figure 11.2 : BOD_5 ***removal and loading for horizontal flow reed beds ($y = 43.7\,x^{0.82}$, $R^2 = 0.75$).***

Research conducted by the University of Birmingham into the use of reed beds for treating farm dirty waters commenced in 1986. Work has been carried out on full-scale red beds on two farms in conjunction with laboratory work at the University. The work has progressed through three separate experimental periods, with the combined horizontal and downflow reed bed systems at the two farms being identical in form, though differing in size. The trials were conducted under farm conditions with varying influents and strengths. The dirty water used in the trials included septic tank sewage effluent, farmyard run-off from heaps of pig manure and dairy parlour washings. The averaged results for the horizontal flow beds for each experimental period are combined in, which presents BOD removal against inlet BOD loading. These results represent over 128 separate analyses with inleet BOD values ranging from 1884 down to 88 mg l^{-1}. Table 11.5 gives an indication of the level of pollutant removal in horizontal flow reed beds.

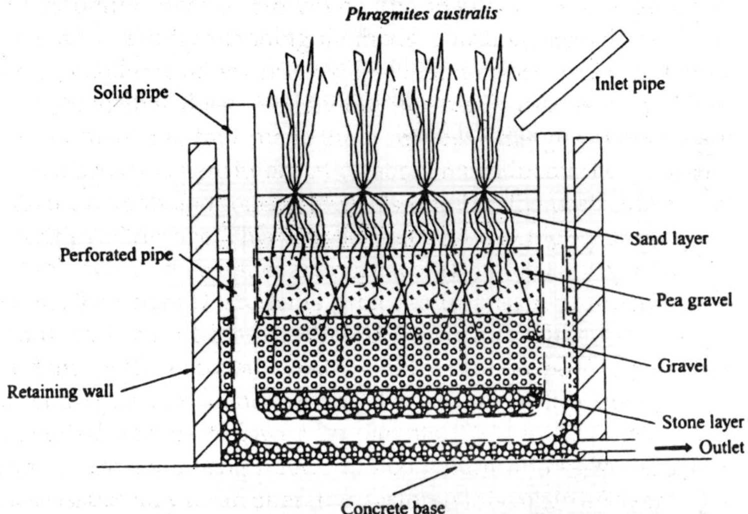

Figure 11.3 : ***Downflow reed bed with Phragmites australis planted in a multilayered matrix of various-sized aggregates. A system of perforated pipes allows countercurrent diffusion of oxygen into the bed and carbon dioxide out of it. The wastewater is flushed onto the bed in pulses large enough to allow even distribution over the bed top surface. These beds are often built above ground between retaining walls.***

DOWNFLOW BEDS

Downflow beds, are constructed of sharp sand, pea gravel, graved and stones in layers of appropriate thickness. Oxygen from the air can diffuse into the bed via perforated pipes set within the layers, as well as from the roots of the reeds. According to the strength of the initial wastewater one, two or three stages of downflow treatment may be required; each stage consists of several identical beds, perhaps six in the first stage, four in the second and three in the third. These undergo alternating operating and resting periods. During the operating period of several days the surface of the sand top layer gradually becomes choked with fine filtered solids leading to flooding conditions; by resting the bed for about a week the soilds are oxidized off and the bed becomes permeable again. Downflow beds gave considerably more potential for oxygenation than have horizontal flow ones. Being self-draining, the residence time in them is short, only a matter of minuts. So far these beds have been designed essentially on hydraulic considerations, to handle the water flow. Experience is slowly being acquired in designing these beds to meet organic loads in terms of kg BOD/day.

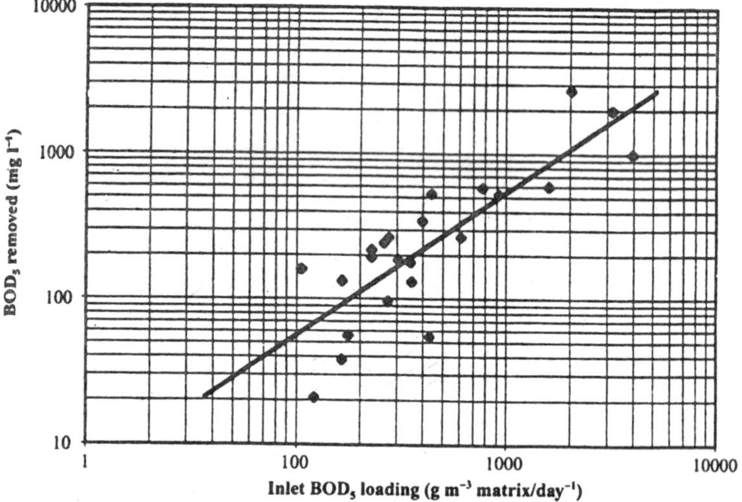

Figure 11.4 : BOD_5 ***removal and loading for*** downflow reedbeds ($y = 0.487x^{1.02}$, $R^2 = 0.71$).

The University of Birmingham research on the farm sites has accumulated information on the performance of downflow beds. This work has shown that some six stages of downflow beds are needed to reduce an influent BOD of 5000 mg l^{-1} down to below 50 mg l^{-1}, the level at which nitrification becomes appreciable. An indication of removal efficiency for various forms of reed bed is given in Table 11.5.

TABLE 11.5 : POLLUTANT REMOVAL BY REED BEDS

Bed	Removal	References
Horozontal subsurface flow	Results from a large number of beds in Europe indicate that BOE removal is 80–90 %, with a typical outlet concentration of 20 gm 1^{-1}, total nitrogen removal is only 20–30 % and total phosphorus removal is 30–40 %	Cooper (1990), Cooper *et al.* (1996)
Downflow	Using two downflow stages to treat domestic sewage achieved removals of domestic sewage achieved removals of 93 % BOD, 90 % suspended solids, 75 % NH_4-N and 37 % orthophosphate	Burka and Lawrence (1990)
Overland flow	Passing the effluent from a lead–zinc mine through a bed of *Typha latifolia* achieved reductions of suspended solids of 99 %, lead of 95 % and zinc 80 %	Lan *et al.* (1990)

BOD: biochemical oxygen demand

Overland Flow Beds

Overland flow beds aim to precipitate metal salts from solution by pH change and aeration; the precipitates are then adsorbed onto biomass on the plant stems or settle onto the matrix surface. Sub-surface flow is impossible in these situations because the precipitates would soon choke the bed. These beds are mainly designed on residence time considerations and can achieve substantial reductions.

Sludge Treatment Beds

Sludge treatment beds comprise a sand/gravel matrix planted with reeds onto which the sludge/slurry is poured periodically, about once a fortnight. Dewatering takes place be drainage and evapotranspiration. If the sludge is carbonaceous in content then oxidation and stabilization can occur. The beds are sized essentially on residence time. As the thickness of solids builds up on the bed the reed and its roots grown up with it. The sludge solids are allowed to accumulate for several years before they are removed; this is easiest done when the reeds have died down in winter.

APPLICATION OF REED BED SYSTEMS

The advantages of constructed wetlands systems include an initial capital cost, which is normally cheaper than that of its mechanical ounterpart. For instance, when a combination of septic tank and horizontal flow reed bed is used for secondary sewage treatment in small rural works, Green and Upton (1993) report a capital cost in te range £ 700–1600 per head of population. This range of costs relates to the size of the unit and to site specific factors. These costs compare with ones of about £2000 per head of population for the alternative mechanized rotating biological contactor installation.

TABLE 11.6 : APPLICATIONS OF REED BED SYSTEMS.

Wastewater type	*Application*
Municipal sewage	Secondary and tertiary treatment
Sewage treatment following septic tanks	Houses, hotels. stately homes, caravan parks. leisure centres, golf ourses

Run-off	From roads, airports and vehicle servicing areas (trains and coaches)
Farm	'Dirty water' from dairy units, slaughterhouses and vegetable washing water
Landfill	Leachate
Industrial	From chemicals manufacture, oil refineries, food and beverage manufacture, paper mills, tanneries
Mine	With metal contamination from operating and derelict coal, metal mines
Sludges, particularly organic ones	Dewatering and stabilization

Where the lie of the land is suitable, gravity flow of wastewater is possible and an electricity supply is not needed. Maintenance and supervision requirements are also low, hence operating costs are for less than those of mechanized units. Mainly because of these cost advantages, reed beds are being constructed in significant numbers for the treatment of weak and medium-strength wastewaters.

However, in addition to economic advantages reed bed treatment systems are aesthetically pleasing and provide wildlife habitats for birds and invertebrates, particularly where they are carefully landscaped and include pools.

With this new technology, new applications are still being attempted and problems exposed. Reliable date on flows is often lacking and flows can vary widely in some instances owning to rainfall. The calcualtion of bed sized has tended therefore to err heavily on the side of caution; this also helps the bed to cope with occasional overload conditions.

Sewage and farm wastes normally provide adequate plant nutrients for growth of the reeds but with many industrial effluents these nutrients are lacking and need to be added to the inflow.

Reed beds are proving adequate for handling wastewaters with BOD levels up to about 500 mg l^{-1}, using a combination of downflow and horizontal flow beds. They have not yet been proven for wastes with higher BOD levels, such as farm 'dirty water' which has a BOD up to about 3000 mg l^{-1}. With farm wastes and many other effluents the concentration of ammoniacal nitrogen is also high, often up to 500 gm l^{-1}. This is proving difficult to break down in simple reed bed systems using the normal processes of nitrification/denitrification and is holding up application of wetlands to such wastes. The evidence gained from horizontal flow beds indicates that, although BOD removal is normally in the range 80–90 %, the removal of total nitrogen can be as low as 20–30 %. After solving the problems of nitrogen removal those of phosphorus removal will become important as this element is heavily implicated in the eutrophication of water masses.

CONCLUSIONS

The technology engineered reed beds is still relatively new. Their application in horizontal flow beds to the tertiary treatment of sewage has proved technically and economically sound in small rural works. The use of downflow beds in supervised situations for slightly more demanding effluents is showing merit. However, their application to strong wastewaters with BODs up to 3000 mg l^{-1} has not yet been resolved and the removal of ammoniacal nitrogen to satisfactory levels is a major drawback to their employment in farm waste treatment. Use of reed beds for metals removal has great potential and will be greatly needed to cope with seepage from abandoned coal and metal mines. Sludge stabilization and drying on reed beds is starting to make progress. Overall there is still enormous potential for advances in reed bed technology.

12

Waste Water Treatment

Rapid industrialization and urbanization over the past decades have generated increasing amounts of wastewater, resulting in environmental deteriotration and pressure on reliable water sources in many countries. Treatment of these waste flows prior to disposal, there fore, is of urgent concern worldwide. In countries with a high gross national product (GNP) substantial investments were made over a period of several decades in high-cost sewerage and centralized treatment facilities. These investments have contributed to a reduction of the waste load to receiving water bodies. However, few of the treatment systems installed have a capacity to remove nitrogen and phosphorus from the effluent, This, together with the increased use of fertilizer over the past 30 years, is frequently causing eutrophication of surface water and groundwater contamination. The situation in countries with a low GNP is even more pressing. Developing nations house the largest part of the world population, take a 90% share of the world-wide population growth, and show strong urbanization patterns. There is at present hardly any infrastructure for the effective treatment of wastewaters in these countries. Most developing countries have only recently nitiated the development of environmental legislation regarding effluent standards and it is expected that the effective implementation of such legislation will need substantial efforts and investments in the coming decades.

Although industrialized nations may have the economical capacity to deal with environmental problems via high-cost technologies, these may not provide an immediate solution for countries with a low GNP. Briscoe (1993) stated that due to the huge cost associated with centralized urban sewerage in Latin America just 2% of the sewage flows are at present properly treated. Grau (1994) calculated that the period of time required to meet EU effluent standards in several central eastern European countries sewers. The challenge ahead, therefore, is to develop reliable and appropriate technology that is within the economical and technological capabilities of developing countries. The non-availability of such technology has in fact been one of the reasons for the limited application of wastewater treatment in many urban and rural situations.

TABLE 12.1 NITROGEN AND PHOSPHORUS UPTAKE BY FLOATING MACROPHYTES

Location	Macrophyte	Daily uptake ($g\ m^{-2}$ per day) N	P	Reference
Florida, USA	Water hyacinth	1.30 (0.25)*	0.24 (0.05)	Reddy and DeBusk, 1987
Florida, USA	Water lettuce	0.99 (0.26)	0.22 (0.07)	Reddy and DeBusk, 1987
Florida, USA	Pennywort	0.37 (0.37)	0.09 (0.08)	Reddy and DeBusk, 1987
USA	*Lemna sp.*	1.67	0.22	Zirschky and Reed, 1988
India	*Lemna sp.*	0.50–0.59	0.14–0.30	Tripathi *et al.*, 1991
Louisiana, USA	Duckwed	0.47	0.16	Culley and Meyers, 1980
Bangladesh	*Spirodela polyrrhiza*	0.26	0.05	Alaerts *et al.*, 1996

* Values in parentheses were obtained during winter season.

This chapter will discuss the role of duckweed-based wastewater treatment in an integrated concept of recycling and re-use of valuable waste components. The focus on duckweed as a key step in waste recycling is due to the fact that it forms the central unit of a recycling engine driven by photosynthesis and therefore the process is sustainable, energy efficient, cost efficient and applicable under a wide variety of rural and urban conditions.

Role of Aquatic Macrophytes

The objective of wastewater treatment is to remove or convert contaminants that are considered detrimental to human health or to the environment. Historically, attention has focused on removal of suspended solids, organic matter and pathogens, but more recently also toxic compounds and nutrient removal are considered to be important issues. Aquatic macrophytes have been studied for their use as effective scavengers of nutrients from wastewaters. Their use has been suggested as a low-cost option for the purification of wastewater and simultaneous production of plant biomass. When applying aquatic plants in shallow ponds a combination of secondary and tertiary treatment may be realized. In addition, aquatic macrophytes assimilate nutrients into a high-quality biomass that may have an economic value. This contrasts with advanced costly nitrification–denitrification, where nitrogen is converted into atmospheric N_2 and therefore is 'lost' for further reuse. This is rather inefficient and a technology based on nutrient reuse seems to be a more rewarding option from both an energy and resource efficiency point of view.

Various studies have reported the use of water hyacinth (*Eichhornia crassipes*), pennywort (*Hydrocotyle umbellata*), water lettuce (*Pistia stratiotes)* and duckweed (*Lemnaceae)* for the efficient removal of nutrients. The economic potential of each plant species for wastewater treatment depends largely on its efficiency to remove nutrients under a wide range of environmental conditions, its growth and maintenance in a treatment system, and the possible application of plant biomass. Water hyacinth has been used most widely, due to its high nutrient uptake capability, but no economically attractive application of the generated plant biomass has been identified so far.

CHARACTERISTICS OF DUCKWEEDS

Duckweeds, or *Lemnaceae*, are small floating aquatic plants, which can be found world wide under widely varying climatic and environmental conditions. The family consists of four genera (*Wolffiella, Wolffia, Lemna, Spirodela*) and at least 37 species have been identified. The plant morphology is simple and structural components are limited to short roots and a frond with a size of only a few millimetres for most species. Parameters affecting duckweed growth in natural environment include temperatrure, light intensity, other climatic conditions, pH and availability of nutrients in the water. These factors will also the important for duckweed growth on sewage, but in this case also the potential of growth inhibition due to high concentrations of sulphides and ammonia need to be considered. Optimum values for the above-mentioned climatic and environmental parameters will differ not only between species but most probably also between varieties of the same species. It is therefore important to use locally available duckweed varieties, since these may prove to give better performance under local conditions.

Growth Rate and Composition

Since structural components like stems and leaves are missing, most of the tissue of duckweed plants is actively involved in photosynthesis. It is not surprising, therefore, that *Lemnaceae* are considered to be among the most vigorously growing plant species on Earth. Duckweed yields reported in literature vary significantly. Low productivity, of 11 t dry matter (DM)/ha/year, was observed by Edwards (1987) when growing *Lemna perpusilla* on septic tank effluent, while values as high as 55 t DM/ha/year were reported by Oron *et al.* (1986) for the growth of *Spirodela polyrrhiza* on domestic wastewater. Highest production rates are usually observed in trials at laboratory scale or small pilot scale. Averge biomass yields in full scale ponds will probable by in the range of 15–30 t DM/ha/year.

Protein content of duckweed depends largely on growth rate and growith conditions. Culley and Epps (1973) reported protein content of 14–26% in duckweed growing on natural water bodies, whereas 29-41% protein was observed in duckweed cultivated in waste stabilization ponds. Duckweed has an enormous potential in protein production, far higher than that of any other crop

(Table 12.2). The relatively high protein content of duckweed is generating interest in animal feed applications of this aquatic plant. *Wolffia arrhiza* is even used for human nutrition in Burma. Laos and northern Thailand. Annual yields of about 2000 kg ha^{-1} have been obtained from very simple cultivation methods in ponds without fertilizer being applied.

TABLE 12.2 ANNUAL PRODUCTIVITY OF DUCKWEED AND SELECTED CROPS IN TERMS OF DRY MATTER AND PROTEIN

Plant/crop	Production	Crude protein	Protein production
	(t DM/ha/year)	(% dry matter)	(kg/ha/year)
Duckweed	17.60	37	6510
Soybean	1.59	41.7	660
Cottonseed	0.76	24.9	190
Peanuts	1.60–3.12	23.6	380–740
Alfalfa hay	4.37–15.69	15.9–17.0	690–2670

DUCKWEED AND DOMESTIC WASTE WATER TREATMENT

Advantages of Duckweed

The use of duckweed for the efficient and low–cost treatment of domestic wastewaters in rural or urban contexts is recommended by various authors because of a number of advantages this technology offers over conventional treatment technology:

- Duckweed grows rapidly and shown a high nutrient uptake rate when exposed to sewage. Duckweed growth is less sensitive to low sewage temperature high nutrient levels, pH fluctuations, and pest and diseases than most other aquatic plants.
- Duckweed can tolerate the rather high concentrations of detergent present in domestic wastewater.
- Duckweed is capable of absorbing, accumulating and/or disintegrating a wide variety of substances, which are not easily

biodegraded in conventional wastewater treatment plants. Accumulation of heavy metals may be disad-vantageous when aiming at feed applications of duckweed biomass.

- Because of its high protein and relatively low fibre content duckweed is highly nutritious and digestible for a wide variety of animals including pigs, ruminants, poultry and fish.
- Harvesting of duckweed plants from the water surface is far less complicated than the harvesting of other macrophytes, which are commonly interconnected over large distances.
- A complete duckueed cover on the wastewater may effectively prevent the development of algae in the water body and provides quiescent conditions, which contribute to a more clear effluent with lower total suspended solids (TSS).
- The presence of a duckweed cover has been reported to decrease the development of mosquitoes in the water body because it prevents mosquito breeding.
- Duckweed positively affects the reduction of the release of odorous compounds.
- In (semi-) arid regions of the world, which are increasingly confronted with water shortage, it is of great relevance to know that water losses due to evapo (transpi-) ration rates can be reduced when covering a water body with duckweed.

These advantages have triggered duckweed research and application. Several full-scale applications of duckweed-based wastewater treatment systems exist, for example in the USA, Bangladesh and China. Duckweed systems have been studied for dairy waste lagoons, domestic sewage, secondary effluent, waste stabilization pond effluents, and fish culture systems. Even in moderate climatic conditions duckweed-based systems are being advocated for treatment of domestic wastewater form small communities in Belgium and Poland. In Siberia the USA-based Lemna Corporation is currently installing a system with an estimated cost of US$50 million, whereas a $30 million agreement was recently signed with China.

Major Treatment Parameters Affected by Duckweed

Aquatic plants used in wastewater treatment are generally con-

TABLE 12.3 DUCKWEED PRODUCTION AND PROTEIN CONTENT

Species	Cultivation	Production (DM/ha/year)	Protein (% DM)	References
S. polyrrhiza	Domestic sewage	17–32	–	Alaerts *et al.* (1996)
L. minor	UASB – effluent	10.7	28.9	Weller and Vroon (1995)
L. gibba	Pretreated sewage	55	30	Oron (1994)
L. gibba	Domestic sewage	10.9–54.8	30–40	Oron *et al.* (1986)
S. polyrrhiza, *L. perpusilla* and W. arrhiza	Septage from septic tank	9.2–21.4	24–28	Edwards *et al.* (1992)
Lemna spp.	Domestic sewage	27	37	Zirschky and Reed (1988)
S. polyrrhiza	Domestic sewage	17.6–315	30	Gijzen (1996)

sidered to not contribute directly to biodegradation processes. Their role is more of a supportive nature, providing surface area for attached biofilms, providing oxygen for aerobic metabolism, and stimulating quiescent conditions in the water phase which enhance settling processes. With more information becoming available, the supporting role of macrophytes and their contribution to overall treatment efficiency is increasingly being recognized.

Nutrients

Lemna species show high nutrient uptake rates in domestic sewage, characterized by abundance in the macronutrients nitrogen, phosphorus and potassium (N, P and K). Typical daily nitrogen and phosphorus uptake rates for duckweed are 0.4 g N m^{-2} and 0.03 g P m^{-2}. Culley *et al.* (1981) calculated daily nutrient uptake rates per m^2 by *Lemnaceae* to amount to 0.4 g N, 0.1 g P, and 0.1 g K. The results from different studies are not comparable, because different species have been used and different climatic and operational conditions applied.

Daily nitrogen removal via microbiological nitrification–denitrification rates may be in excess of 1 g m^{-2}. Culley *et al.* (1978) demonstrated a 20–40% lower nitrogen content in pond effluents covered with *Lemnaceae* than in algal-based pond effluents without aquatic plants.

Pathogen Removal

Pathogen removal from duckseed-covered lagoons treating domestic wastewater is important in order produce an effluent quality that allows for further reuse in agriculture and/or aquaculture. Therefore the effluent guidelines of < 1000 coliforms/100 ml and helminth eggs < 1/1 should be met to ensure safe effluent reuse. Other health-related concerns are related to the manual harvesting of duckweed and its application as animal feed. These cocerns need to be studied and evaluated carefully because they may directly affect the feasibility of duckweed application to domestic wastewater treatment.

Islam *et al.* (1996) monitored faecal coliform levels in a plug-flow duckweed lagoon and found a reduction from 4.5×10^4 per

100 ml in the influent to values below 100 per 100 ml in the effluent. These results are very positive, but do not provide information on the relative contribution of duckweed on coliform removal or survival. Dewedar and Bahgat (1995) reported that the decay rate observed for faecal coliforms exposed to direct sunlight was 0.177 h^{-1}, whereas faecal coliform numbers under a dense cover of *Lemna gibba* did not decline over a period of 5 days. These results suggest that elimination of exposure to direct sunlight by a duckweed cover may result in extended survival for faecal coliforms in treatment ponds. In addition, Culley and Epps (1973) showed that the presence of a duckweed cover reduces the dissolved oxygen level in the water phase and thus diminishes the disinfection effects of free oxygen radicals produced in the water phase.

Efficient removal of coliforms has been reported for conventional algal-based pond systems. Direct effects of sunlight penetration, combined with indirect stimulation of algal growth resulting in fluctuations of pH and dissolved oxygen, stimulate coliform reduction. Van Haandel and Catunda (1997) demonstrated a positive correlation between pond depth decay rate for *Escherichia coli*, stressing the effect of sunlight penetration into the water phase. With deeper ponds and presence of a duckweed cover disinfection via solar radiation will decrease and coliform die-off rates will be reduced. Pathogenic parasites usually settle in ponds. Duckweed might enhance this process by improving the hydraulic stability of the water column.

BOD Removal

The role of macrophytes in biochemical oxygen demand (BOD) removal is not fully understood and probably varies between different species. The introduction of duckweed to a pond limits the free diffusion of oxygen from the air into the water phase. A compensating effect may be the diffusion of produced oxygen through the duckweed roots. Oxygen will induce aerobic mineralization of organic compounds by the heterotrophic microbial biomass adhering to the plant surface. In this concept, ponds with a dense cover of aquatic plants could be considered as fixed film reactors.

Hillman (1961) reported that duckweed may contribute to the removal of simple organic compounds via direct uptake. Korner *et al.* (1998) could not confirm this, but suggested that aerobic conditions at microlevel in the root zone might induce high BOD conversion. Zirschky and Reed (1988) reported that duckweed supports attached biofilm development to the roots, but its quantitative contribution to overall BOD removal is not clear. Alaerts *et al.* (1996) reported BOD removal efficiencies of 95–99% in a 0.7 ha sewage lagoon which was operated with a cover of *S. polyrrhiza* at a hydraulic retention time of about 20 days. Daily BOD loading rate in the lagoon varied between 48 and 60 kg ha^{-1}. DeBusk and Reddy (1987) observed BOD removal rates as high as 300–400 kg/ha/day under very high BOD loading conditions in treatment systems with water hyacinth. However, such highly loaded systems usually produce an effluent BOD level of over 60 mg^{-1}.

Rather than looking for ways to enhance BOD removal in duckweed ponds it is much more rewarding to consider the use of anaerobic technology before duckweed ponds. Via anaerobic treatment technology the organic matter is reduced by 65–85%. The resulting effluent is effectively stripped from TSS and BOD, but has conserved all other contaminants such as nitrogen, phosphorus, and faecal coliforms. The application of post treatment by macrophyte-based ponds seems therefore most logical and feasible.

TSS Removal

The removal of TSS in waste stabilization ponds is mainly determined by the rate of organic particle biodegradation, particle settling rates, and the potential production of algal-related TSS. In particular, in developing countries high concentrations of TSS are encountered in the raw sewage, which might enhance flocculent settling and contribute to substantial BOD and TSS removal in anaerobic ponds or reactors. The development of algae results in an increase of effluent TSS to over 200 mg 1^{-1}. Since algal growth will be largely suppressed in duckweed-covered ponds TSS removal will be efficient and allow effluent distibution to agriculture via drip irrigation systems without rapid emitter fouling.

Removal of Heavy Metals

Heavy metal removal mechanisms in macrophyte-based treatment systems include plant uptake, chemical precipitation and adsorption. *Lemnaceae* show a fair degree of tolerance to heavy metals and posses a great ability to accumulate them. Landholt and Kandeler (1987) suggested that duckweed could effectively be used to remove metals from wastewater flows. Gaur *et al.* (1994) found that the accumulation of cadmium, chromium, cobalt, copper, nickel, lead and zinc by *S. polyrrhiza* and *Azolla pinnata* was related to the concentration of metals applied.

The accumulation factor for heavy metals in *Lemnaceae* depends largely on its concentration, the presence of other metals, and the duckweed species. For lead and cadmium accumulation factors of over 1000 have been reported. *Lemna valdiviana* accumulates copper by a factor of 500–54 000, depending on concentration offered, whereas *Lemna minor* showed accumulation factors from 80–8000. Extremely high accumulation factors of 660 000 and 850 000 were reported for aluminium and manganese, respectively Gellini and Piccardi (1981) reported that *L. minor* causes a 75% reduction in copper content of water containing 5 mg l^{-1} within 48 h. This suggests that duckweed can be effectively used for heavy metal stripping. However, the duckweed produced under these conditions should be considered as a chemical waste, and not be used in any feed application.

Duckweed Ponds in Combination with Anaerobic Technology

The application of anaerobic technology results in a substantial reduction of organic matter and suspended solids from wastewater. In cases where sewage is concentrated and has temperatures above 15°C, anaerobic technology can be most rewarding. The technology does not require complicated and costly mechanical equipment or intensive energy input. On the contrary, recovery and valorization of the produced high caloric methane gas may positively affect the overall economy of the process. Anaerobic technology show very poor nitrogen and phosphorus removal rates, whereas typical removal rates for coliforms via attachment to and settling of suspended solids is only some 50–90%. Helminth eggs are more susceptible to removal by filtration, entrapment and settling within the anaerobic reactor. As duckweed ponds focus on the reduction of pathogens and dissolved nitrogen and phosphorus compounds, a combination with anaerobic pretreatment seems to be most effective. Advantages of such a combination are:

- The design of duckweed ponds can entirely be based on nutrient removal and pathogen reduction rather than on the biodegradation of organic matter and the subsequent transfer of substantial amounts of oxygen per m^2 per day. This may lead to a significant reduction of HRT and pond area required to achieve effective treatment results.
- Accelerated nutrient uptake rates can be achieved if substantial chemical oxygen demand (COD) reduction is realized in the anaerobic reactor prior to the duckweed pond.
- Anaerobic pretreatment contributes to the liquefaction of suspended organic particles. In this process organically bound nutrients will be mineralized and will become available for duckweed growth.
- Anaerobic technology as well as duckweed pond technology can be easily scaled down to small capacity and can thus be applied to decentralized sewage treatment in urban areas in developing countries, or to small communities world-wide.

In particular for highly concentrated wastewaters from the agro-industrial sector anaerobic treatment is economically feasible. Breweries, slaughterhouses, dairies and other food industries already make use of this technology in order to treat their effluents. As long as no strict criferia are imposed on effluents via legislation there might be no driving force to add post treatment technology to the anaerobic reactors. However, it could be an economic incentive to initiate the application of duckweed pond technology to these effluents on a commercial basis in those areas where duckweed has economic potential. In the rural setting it will be interesting to check whether duck weed technology can successfully be applied to anaerobically digested manure and thus contribute to the recycling of energy and nutrients.

Other Sources of Waste

Food Processing Industry

Almost no information is available regarding the treatment of specific industrial wastewater in duckweed ponds. Duckweed may be successfully applied for the treatment of nondomestic wastewater with a high nutrient content such as effluents from food-processing businessses, slaughterhouses and chemical industries involved in the production of nitrogen and phosphorus compounds. Wastewaters with extremely high nitrogen levels in combination with high pH may be toxic to duckweed. Generally no inhibitory effects are observed at concentrations up to 50 ppm ammonium at pH levels below 8. The sensitivity to ammonium may differ for different duckweed species and may be decreased by gradual adaptation to higher ammonium concentrations.

Algal Culture Effluent

Koles *et al.* (1987) demonstrated that duckweed can be used for the efficient removal of ammonia, phosphorus and TSS in the effluent from and algae culture unit. This application is unique, since the effluent has a high pH and contains high concentrations of phosphorus and ammonium ions. The authors reported and average duckweed production of 87 g fresh weight m^{-2}/day, with a maximum value of 280 g m^{-2}/day (about 50 t DM ha^{-1}.year). Another possible application of duckweed has been reported for the treatment and recycling of water from intensive fish cultivation.

Xenobiotics

The removal of xenobiotic compounds such as phenol, naphthalene and chlorodibromomethane from wastewater by a water hyacinth-covered system in a pilotscale study has been demonstrated by Conn and Langworthy (1094). Federle and Schwab (1989) reported the efficient biodegradation of alcohol ethoxylate and mixed amino acids by the microbiota associated with the duckweed *L. minor*. The results of these studies indicate that macrophytes do accelerate the biodegradation of xenobiotics, but the mechanisms involved need to be further studied.

Manure

In the rural context, other sources of waste materials can be considered and their treatment in duckweed ponds could contribute to the recycling of valuable components. In principle, any organic meterial that is readily biodegradable and has a high nutrient content could be treated in duckweed ponds. Examples of such waste materials include all kinds of animal manure, latrine wasters, effluents from biogas plants, kitchen waste, composted agricultural wastes and market wastes. Treatment of latrine septage in a duckweed system could provide basic sanitation and simultaneously stimulate nutrient recycling in the rural areas of low GNP countries. At the same time, the duckweed ponds could be fed with other waste streams, such as digested manure and composted agricultural and domestic wastes.

INTEGRATED CONCEPTS

Integrated systems have been defined as a combination of processes and practices where optimum use of resources is achieved via waste recycling aimed at the recovery and reuse of energy, nutrients and possibly other components. The conversion processes for different source of waste are arranged in such a way that a miniumum input of external energy and raw materials is required and maximum self-sufficiency is achieved. In rural Asia, integrated systems form an old concept that has been applied for hundreds or even thousands of years. In China there are huge farms, which are almost completely self-sufficient in terms of energy and nutrients because of effective recycling of their waste streams. The application of integrated concepts provides a good balance between resource utilization, reuse and environmental protection. Such balance is completely absent in denely populated urban areas, with high consumption rates and where concentrated waste production is sustained by the large-scale importation of energy and nutrients into the urban environment. The absence of reuse processes for energy, nutrients and other valuable components in urban waste has generated serious environmental and public health concerns. It is suggested therefore that urgent attention be given to the development of rational reuse strategies in the urban context.

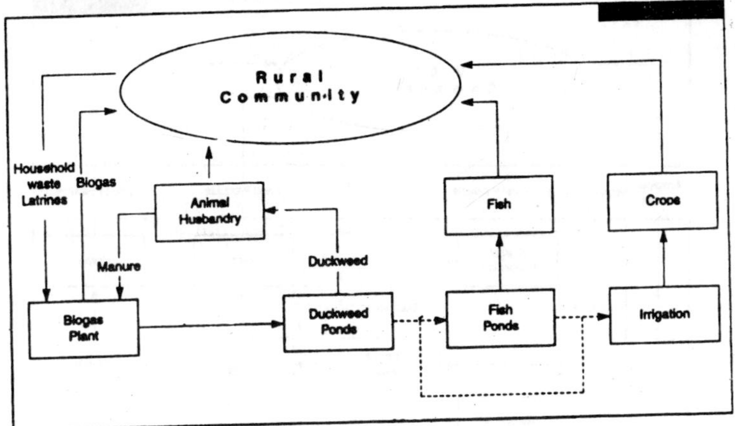

Figure 12.1 : ***Community-based rural integrated waste recycling system using duckweed ponds.***

This approach, in combination with waste minimization schemes, will provide a more sustainable solution than the present approach of costly end of pipe treatment. The challenge for the coming years therefore will be to develop integrated concepts and processes for the minimization, recovery and reuse of waste materials in both rural and urban environments in high and low GNP countries. If effective programmes and action plans could be defined by relevent organizations, such as the World Bank. United Nations, national universities and research centres, this challenge could be met within a reasonable time with the help of nodern science and technology.

Duckweed ponds could provide an important role in recycling and reuse schemes in both rural and urban areas. The process steps and products of an integrated duckweed based treatment system for rural and urban recycling of waste streams are presented. Initially anaerobic technology is advocated to reduce the bulk of organic and suspended matter. The energy produced, either in rural biogas digesters or urban high-rate reactors (e.g. upflow anaerobic sludge blanket; UASB), can be used by the community (rural context) or for the operation of subsequent treatment steps (urban application), thereby reducing treatment costs. Various full-scale anaerobic reactors have been installed to treat domestic wastewater in India, Colobmia, and Brazil.

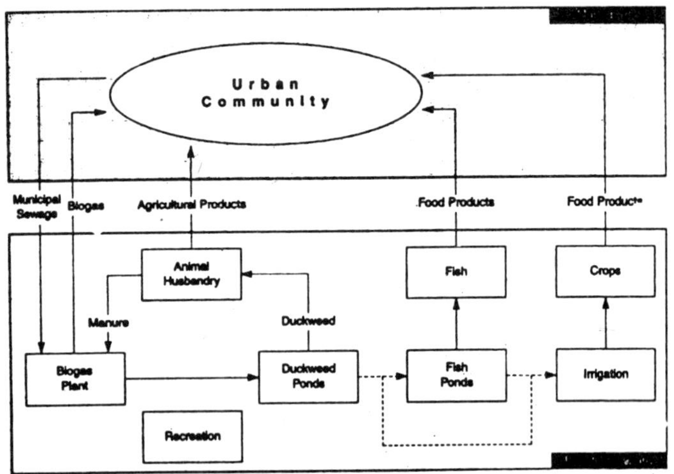

Figure 12.2 : *Urban integrated waste recycling system using duckweed ponds.*

Duckweed Production and Nutritive Value

Various authors recommend the application of duckweed biomass as a source of high quality protein in animal nutrition. The advantages that make duckweed particularly attractive as an economic biomass resource were listed. Besides these positive characteristics, there are also a numer of negative factors that need to be studied when considering duckweed as a potential resource:

- Owing to the efficient absorption of heavy metals and possibly other toxic compounds, duckweed should be cultivated using effluents containing low concentrations of such compounds. No information is available on the possible transmission of pathogens if duckweed harvested from domestic wastewater treatment ponds is used as an animal feed.

TABLE 12.4 FEED CONVERSION RATIOS FOR DUCKWEED TO FISH

Duckweed species	Fish species	FCR	Reference
Lemna	Tilapia	1.6–3.3	Hassan and Edwards,

1992			
Unknown	Grass carp 3 g	1.6	Shireman *et al*., 1978
Unknown	Grass carp 63 g	2.7	Shireman *et al*., 1978
Uuknown	Uunknown	1.6–4.1	Baur and Buck, 1980
Unknown	Unknown	3.1	Hajra and Tripathy,
1985			
Spriodela	Polyculture	1.2–3.3	PRISM, Bangladesh

- Duckweed has a relatively high moisture content (about 95%), which will affect the cost of handling, transportation and drying. This characteristic will be less important in an integrated system, where duckweed is used on site.
- The genera *Lemna* and *Spirodela* may contain high amounts of calcium oxalate. The presence of this component may limit the use of certain duckweed species for non-ruminant and human nutrition. When hit is properly mixed with other feed, constituents no harmful effects are expected.
- Production of the crop is limited to areas with a minimum of 20 cm of standing water throughout the year or during a major part of the year.

Biomass productivity and nutritive value are important factors when considering duckweed as an animal feed. The nutritive value largely depends on the type of animal, digestibility of major plant components, and protein content. Frye and Culley (1981) reported a lower limit of 20–30 mg nitrogen per litre in the medium in order to achieve high protein content in duckweed biomass.

Duckweed as an Animal Feed

Most feeding trials with duckweed have been done with fish cultivation. The *Lemnaceae* appear to be suitable for a wide range of fish species and may also be applied in polycultures of fish.

Duckweed may be fed fresh as a sole feed or in combination with other feed components. In addition, can be dried and offered in pellets. Intensive fish production with duckweed as a predominant feed constituent has been reported by a number of authors. Promising results have been obtained with the grass carp *Ctenopharyngodon idella*. The feed conversion ratio (FCR = g DM duckweed per gram fish fresh weight) is lower in young grass carps than in mature fish. The reported FCR values for duckweed compare very favourably with FCR values reported for other fish feeds such as catfish chow or rye grass.

PRISM, a local NGO in Bangladesh, has experimented with duckweed-based wastewater treatment and fish production for more than 9 years. The ponds yield over 12 tons ha^{-1}/year of fish, making the overall treatment process profitable with an estimated net annual revenue of US$3000 ha^{-1}. Excellent results have been obtained for both duckweed and polyculture fish production. *Lemnaceae* may also be used for shrimp cultivation or to grow zooplankton, which subsequently is used by carnivore fish species.

For higher animals (ruminants, pigs and chicken) preliminary trials show that small portions of duckweed in the diet stimulate their growth, whereas high proportions of duckweed tend to decrease weight gain or create digestion problems.

Effluent Reuse

In view of the rapidly growing shortage of renewable water resources in many parts of the world there is a growing interest in the use of treated effluents from wastewater treatment plants. These effluents not only serve as water source but also provide nutritious input for fish and crops in aquaculture or agricultural irrigation schemes. In most developing countries the agricultural sector is by far the largest consumer of water resources (typically over 90%) compared to the domestic and industrial sector. In many countries the conventional 'supply-driven' water consumption has already led to uncontrolled abstraction and substantial groundwater depletion in many arid and semi-arid areas. Reuse of wastewater effluents could definitely provide some relief on the use of scarce water resources. Duckweed ponds could condition the sewage effluent and make it fit for reuse. When looking at the water quality criteria required for reuse of wastewater in agriculture duckweed ponds definitely have good potential for effluent cleansing.

The effluent of duckweed ponds may be useful in drop irrigation shemes applied in arid regions of the world as the duckweed cover helps to minimize water losses, reduces the salinity of the effluent by the uptake of dissolved ionic compounds, and prevents the occurrence of algal blooms. Emitter fouling can thus be minimized as the amount of suspended solids in the irrigation water will be low.

There seems to be a contradiction when promoting the use of sewage nutrients for duckweed growth while claiming at the same time that the final effluent can supply nutrients for crop production. Domestic sewage in developing countries is characterized by high nutrient concentration. In order to meet the crop water and nutrient requirements the concentration of nutrients in the effluent needs to be carefully matched to the water demand, which frequently urges the need for effluent dilution with fresh water. A better option is to partially reduce the nutrient concentration before agricultural application. Another reason for applying duckweed ponds in integrated waste recycling systems is that excessively high nitrogen levels in effluents might be toxic to crops or fish, or create substantial groundwater pollution when recharged into the soil in an uncontrolled way.

Recreation

A major limitation for the application of duckweed technology in urban centres is the non-availability of space at a reasonable cost. However, outside the city, land cost will be substantially lower and duckweed ponds may be feasible. Anaerobic treatment facilities may be planned at covenient locations in or near the city, requiring only limited space. The effluent of the anaerobic reactors can be channelled away from the city to duckweed pond facilities. The duckweed harvested at regular intervals can be used to cultivate fish requirement for the duckweed pond is estimated at 0.5–1 m^2 per capita in (sub-) tropical climatic regions. Although this is singnificantily lower than for conventional stabilization ponds. considerable land area will be required.

Owing to the rapid growth of cities, sufficient land area has to be reserved for recreational purposes. In mega-cities with over 1 million inhabitants large-scale recreational facilities can only be planned outside the city at convenient locations. In recognition of the fact that the green duckweed-covered ponds provide a pleasant ecological appearance, one might think of a multifunctional use of space, combining wastewater management with a recreational destination. The posssible production of unpleasant odours from the wastewater is effectively reduced by the presence of a duckweed cover on the pond surface. This multifunctional use of the available space may greatly enhance the economical feasiblity of the entire system. Together with the income from the products generated (energy, fish, irrigation water), the proposed integrated system has the potential to become a commercial enterprise generating substantial revenues.

CONCLUSIONS

- Conventional treatment technology does not provide a sustainable solution of wasterwater management in countries with relatively low GNP. The objective of developing sustainable solutions for waste treatment could be achieved if waste treatment is considered in an integrated concept together with cleaner production and reuse.
- We can learn from the processes and efficiencies of natural systems how efficiently biomass is constantly produced and recycled. Such natural processes have been optimized over millions of years of evolution and it would therefore be wise to simulate and stimulate these processes when developing waste treatment and reuse technology. The duckweed-based integrated concept as presented in this chapter is basically a simulation of natural processes and systems. During the development and testing of such technology the best results can be expected if (sanitary) engineers work closely together with ecologists, biologists and chemist in

an interdisciplinary approach.

- The combination of duckweed-based wastewater treatment with cultivation of herbivorous fish could be a most rewarding low-cost technology for waste recycling that can be applied to both rural and urban waste and wastewater. In the urban context, such a process could be combined further with a recreational function, which will positively affect its economical feasibility. An integrated approach has the potential to generate net profit from the treatment of waste. The implementation of reuse-oriented treatment systems therefore could be taken up by self-sufficient and even profitable enterprises. The operation of the integrated system will also generate employment, which is a high priority in most countries.
- Anaerobic pretreatment of waste and wastewater may have several important advantages, including production of valuable energy and reduction of land area required for subsequent treatment steps. In combination with duckweed-covered ponds, anaerobic technology further reduces the chances of production of bad odours and therefore offers the possibility of combining the treatment function with other functions (recreation, fish production etc.).
- The technical and economical feasibility of selected integrated treatment systems should be demonstrated via pilot and full-scale demonstration projects at various locations under different conditions of wastewater composition, climate, culture etc. The World Bank, together with other (inter) national actors, could play a most important role in the planning and financing of such initiatives which could induce a world wide breakthrough in waste treatment and reuse.

13

Conserving Plants in Danger

It is estimated that there are more than 270 000 plant species in existence and approximately 34 000 or 1 in 8, of these are considered endangered (IUCN, 1998). Competition with increasing human populations and the resulting loss of habitat are contributing to an increasing rate of plant species extinction. Some researchers have estimated that one quarter of plant species are at risk of extinction within the next generation. Approximately one-fifth of the 20 000 native species in the United States are of concern, and possibly 800 of these may be lost in the next decade unless attention is given to their conservation.

There are several approaches to avoiding these losses. The preservation of species *in situ* is of primary importance for maintaining the broadest range of plant diversity. As a supplement to this, *ex situ* preservation can play a role in backing up taxa which are particularly threatened or are rare in the wild. The maintenance and propagation of species in botanical gardens and arboreta as well as in seed and spore banks has traditionally provided a valuable safeguard against loss for many rare species (Lalierte, 1997).

Modern biotechnologies offer the potential of extending these traditional *ex situ* preservation and propagation methods of an even broader range of taxa and tissues types. These techniques have been developed primarily for agricultural and horticultural species, but are increasingly being applied to collecting, propagating, preserving, and evaluating rare and endangered plant germplasm as well.

ENDANGERED SPECIES

The limited amount of plant material available is an over-riding factor in endangered species work. The ability to test protocols and even to conduct replicated experiments may be severely limited, and generally, work with related, non-endangered species is used as a guide. If only a small number of seeds are produced by a species, the use of non-seed tissues my be advised. In addition plants may be located in remote or difficult-to-access areas and collecting trips may be expensive. Permits are often also required for any collections that are made, as well as for transport and importation.

The resources available for work with endangered species are also often limited, compared with economically important species. In addition, there is often a disparity in the resources allocated to the conservation of plants, in comparison with animals. For example in the United States, only 3 per cent of federal funding for the conservation of endangered species is directed towards plants, even though comprise one third of the total species listed as endangered or threatended in the United States.

Despite these limitations, a significant amount of work has been done with endangered plant species using biotechnologies to solve specific problems in the areas of propagation, germplasm preservation, collection and analysis. These will be described and some examples of their application and of the particular needs of individual species will be given.

IN VITRO PROPAGATION TECHNOLOGIES

Although many endangered plant species can be propagated successfully by seed or by cuttings, there are some species that do not reproduce well by these traditional methods. One of the earliest programmes to use *in vitro* propagation methods for rare and endangered species was the micropropagation unit at the Royal Botanic Gardens, Kew established in 1974. Since that time, there have been several reviews on the use of *in vitro* methods for en-

TABLE 13.1 APPLICATION OF BIOTECHNOLOGY TO THE CONSERVATION OF ENDANGERED PLANT SPECIES

Family	species	Status	Propagation methods	References
Acanthanceae	*Adhatoda beddomei*	Over-coll, med, few seeds, slow prop,	Shoot tips	Sudha and Seerni (1994)
Agavaceae	*Agavè arizonica*	Prop for res or cult	Leaf – cal shoots	Powers and Backhaus (1989)
	Agave victoria-regtinae	Slow prop	IV lf – som emb	Rodriguez-Garay *et al.* (1996)
Aloeaceae	*Haworthai* spp.	Self-sterile, slow prop, hort	Lf – adv sh	Rogers (1993)
	Kniphofia pauciflora	prop ofr cult	Stolon – shoots	McAlister and van Staden (1996)
Alstroemeriaceae	*Leontochir ovallei*	Over-coll, med	IV germ, microprop	Lu *et al.* (1995)
Amaryllidacae	*Crinum macowanii*	Over-coll, med	F1 stem – adv sh	Slabbert *et al.* (1995)
	Gethylis linearis	Rare	Bulb scale – adv bulbs	Drewes and van Stanen (1994)
	Leucojum aestivum	Rare	Lf – adv sh	Stanilova *et al.*
	Narcissus bugei, N. longispathus, N. nevadensis, N. torifolius	Rare, endemic Clemente (1991)	Scale lf – (1994)	
Apiaceae	*Glehnia littoralis*	Rare, med	Rhozome – adv sh	Hui *et al.* (1996)
Apocynaceae	*Rauwolfia micrantha*	Poor germ, poor rooting, med	Sh tips, nodes	Sudha and Seeni (1996)
	Rauwolfia sepentina	Over-coll, med, poor seed viability	Nodes – microprop	sharma and Chandel (1992)
	Wrightia tomentosa	Over-coll,	Nodes	Purohit *et al.* (1994b)
Araliaceae	*Acanthopanax koreanum*	Over-coll, med, slow prop	Stem- callus-somatic embryos	Choi *et al.* (1997)
Aristolochiaceae	*Aristolochia indica*	Over-coll, med,	Sh tip and nodes Lf – adv sh	Manjula *et al.* (1997)
Asclepiadanceae	*Stapelia semota*	Rare	Ax buds	Mohamed-Yasseen *et al.* (1995)
Asteraceae	*Artemisia alba*	Rare, one population	Microprop	Ronse (1990)
	Artemisia granatensis	Few seeds available	Sh tips	Clemente *et al.* (1991)
	Atractylis arbuscula var. *Schizogynophylla*	Rare, endemic	Sh – microprop	Gonzalez Aleman *et al.* (1989)
	Centaurea junoniana	One pop, threatened by lava	IV germ – microprop, adv sh	Hammatt and Evans (1985)
	Leontopodium alpinum	Over-coll	Inflorese – adv sh	Zapratan (1996)
	Leptinella nana	prop for res and reintro	Sh tips (1994)	Carson and Leung
	Olearia microdisca	End endemic	Nodes	Krogstrum and Norgaard (1991)
	Saussurea lappa	Rare, med	Sh tip – microprop	Arora and Bhojwani (1989); Johnson *et al.* (1997)
	Senecio hadrosomus	Rare, endemic, low seed set, damaged by insects, germ rate low	Sh – microprop	Bramwell (1990)
Begoniaceae	*Begonia* spp.	Prop for cult	Lf – adv sh,	Bowes and Curtis (1991)
Betulaceae	*Betula uber*	30 individuals	Buds	Vijayakumar *et al.* (1990)
Boraginaceae	*Hackelia venusta*	Over-coll, hab loss	Sh tips	Edons *et al.* (1996)
Brassicaceae	*Coronopus navasii*	Rare, endemic	IV germ – seedlings – adv sh	iriondo and Perez (1990b)
	Draba aizoides		IV germ to microprop	Ronse (1990)

Table 13.1 contd.

Family	species	Status	Propagation methods	References
	Fibigia triquetra	Rare	IV germ – microprop	Prevalek-Kozlina *et al.* (1997)
Bromeliaceae	*Dyckia macedoi*	Rare, endemic	IV germ – microprop Lf – adv sh	Mercier and Kerbauy (1993)
	Vriesia fosteriania V. hieroglyphica	Over-coll, prop for hort	Seedling – adv sh	Mercier and Kerbauy (1995)
Cactaceae	*Astrophytum Capricorn*	Rare	IV germ – microprop	Cardenas *et al.* (1993)
	Aztekium ritteri	Over-coll, slow growing	IV offshoots	Rdriguez-Garay and Rubluo (1992)
	Escobaria Escouriensis, E. robbinsorum, Mammillaria wrightii, Pediocactus bradyi, P. despainii, P. knowltonii, P. paradinei, P. winkleri, Sclerocactus. mesae-verdae, S. spinosior, Toumeya papyracantha,	Over-coll	shoot tips	Clayton *et al.* (1990)
	mammilaria san-angelensis	Over-coll	IV Germ – ax and adv shoot form	Martinez-Vazquiz and Rubluo (1989)
Campanulaceae	*Nesocodon* subsp. *bohemicus*	Rare, endemic	Sh tips, nodes – microprop	Kovac (1995)
Caryophyllaceae	*Dianthus arenarius* subsp. *bohemicus*	Rar, endemic	Sh tips, nodes – *microprop*	kovac (1995)
	Paronychia chartacea	Few sees and low Viability, hab loss, prop for reintro	IV germ – microprop	McKently and Adams (1994)
Cistaceae	*Helianthemum Polygonoides*	Rare, endemic	IV germ to microprop	iriondo *et al.* (1995)
Combretaceae	*Anogeissus rotundifolia*	Rare, endemic	IV germ – microprop	Singh and Shekhawat (1997)
Cupressaceae	*Juniperus cedrus*	Germ low, rooting diffcult	Embryo – adv sh	Hary *et al.* (1995)
Dioscoreaceae	*Dioscorea balcanica, D. caucasica*	Rare, endemic	Lf and zygotic Somatic embryo	Chulafich *et al.*
	Trichopus zeylanicus	Few seeds, slow seed maturation, hab loss, med	IV germ – microprop	Krishnan *et al.* (1995)
Droseraceae	*Dionaea muscipula*	Prop for hobbysits	Lf – adv sh	Kukulczanka *et al.* (1989a)
	Dionaea inusipula, Drosera spp.	Prop for preserv	Lf – adv sh	kukulczanka (1991)
	Drosera intermedia, D. rotundifolia	Protected species	IV germ – microprop	Ronse (1990)
Epacridaceae	*Leucopogoon* obtectus	One population	sh tips	Bunn *et al.* (1989)
Euphorbiaceae	*Euphorbia handiensis*	Gran Canaria	Shoot culture to microprop	Gonzalez Aleman *et al.* (1988)
Fabaceae	*Glycyrrhiza glabra*	Rare, med	Sh tips – microprop	Dimitrova *et al.* (1994)
	Pterocarpus marsupium	Rare	IV germ – microprop	Das and Chatterjee (1993)
	Sophora toromiro	Extinct in wild	Seedling – nodes	iturriage *et al.* (1994)
	Trifolium stoloniferum	Two populations	Sh tips	singha *et al.* (1988)
Gentianaceae	*Centaurium rigualii*	Rare, endemic	Nodes	iriondo and Perez (1996)

Table 13.1 contd.

Family	species	Status	Propagation methods	References
	Gentiana kurroo	Over-coll, med	Sh tips, nodes	Sharma *et al.* (1993)
	Gentiana lutea	Over-coll	IV germ – microprop	Momcilovic *et al.* (1997)
	Swertia chirata	Rare	Seedling stem – callus – adv sh	Shrestha and johi (1992)
Globulariaceae	*Globularia ascanii*	Rare, endemic	IV germ to microprop	Cabreta-Perez (1995)
Goodeniaceae	*Lechenaultia pulvinaris*	Rare Sh tip –	Rosetto *et al.* (1992) microprop	
Guttiferae	*Kielmayara coriacea*	over-coll, med	IV germ microprop	Arello and Pinto (1993)
Haemodoraceae	*Conostylis wonganensis*	Rare	Sh tip – microprop	Rosetto *et al.* (1992)
Lamiaceae	*Coleus forskohlii*	Over-coll, med	Nodes	Sharma *et al.* (1991)
	Hedeoma multiflorum	over-coll, med	lf – adv sh, Sh, nodes	Koroch *et at.* (11997)
Lauraceae	*Ocotea catharinensis*	Seeds poor germ, brief viability,	Zygotic emb – som emb	Campos and Pais (1993)
	Persea indica	Seed germ low, prop for preservation	IV germ – microprop	Campos and Pais (1996)
Liliaceae	*Allium tuberosum*	Over-coll	Basal plate – adv sh	Radhamani and Chandel (1992)
	Chlorophytum borivilianum	Over-coll, med, seed Germ low	Shoot bases – microprop	Purohit *et al.* (1994a)
	Fritillaria releagris	Prop for reintro and hort	bulb scale	kukulczanka *et al.* (1989b)
	Lilium rhodopaeum	Rare	Bulb – adv sh	Stanilova *et al.* (1994)
Sandersonia	Over-coll		Seeds,embs – callus – shoots, Thubers – shoots	Finnie and van Staden (1989)
	Sowebaea multicaulis	Rare microprop	Sh tip –	Rosetto *et al.* (1992)
	Stawellia demorphantha, Wurbea spp.	Rare	Microprop	Dixon and Keighery (1992)
	Thuranthos basuticum	Over-coll	Bulb scale	Jones *et al.* (1992)
	Trillium persistens	Rare	Stem and lf – adv shoots	Pence and Soukup (1995)
Lythraceae	*Woodfordia friticosa*	Rare, med	Sht tips. nodes	Krishnan and Seeni (1994)
Malvaceae	*lavatera oblongifolia*	Rare, endemic	microprop	iriondo and Perez (1990b)
Marattiaceae	*Angiopteris boivinii*	Rare	IV spore germ	Fay (1992)
Meliaceae	*Turraea laciniata*	Rare, endemic	Sh tips, nodes – microprop	Krogstrum *et al,* (1990)
Myoporaceae	*Eremophila resinosa*	Rare	sh tip – microprop	Rosetto *et al,* (1992)
Myrtaceae	*Eucalyptus graniticola*	Rare	Sh tip – microprop	Rosetto *et al.* (1992)
Nepenthaceae	*Nepenthes khasiana*	Rare, medicinal	IV germ – microprop	Latha and Seeni (1994)
	Nepenthes khasiana	Rare	Sh tips, nodes – microprop	Seeni (1990); Rathore *et al.* (1991)
	Nepenthes pervillei	Rare, prop for reintro	IV germ – microprop	Redwood and Bowling (1990)
Oleaceae	*Forsythia coreana*	Rare	Node – microprop	Lee *et al.* (1995)
Orchidaceae	*Cattleya dowiana*	Rare	IV seed germ	Marlow and Butcher (1987)
	Cypripedium calceolus	Rare	Imm seed germ	Malmgren (1992)
	Cypripedium debile,	Rare	IV germ, Nodes –	Hoshi *et al.* (1994)

Table 13.1 contd.

Family	species	Status	Propagation methods	References
	C. henryi, C. japonicum, C. tibeticum, C. montanum		microprop	
	Cypripedrum reginae	Rare	IV germ	Faletra *et al,* (1997)
	Dendrobium lindleyi (1997)	Rare	IV germ	Kaur and Sarma
	Dendrobium Moniliforme	Rare	Microprop	Lim *et al.* (1993)
	Dendrobium Spectatissimum	Rare	IV germ	Marlow and Butcher (1987)
	Disa uniflora, Eurychone Galeandre, laelia jongheana, Pleione formosana, Sarcochilus fitzgeraldii, S. hartmanii, Sohralia xanotholeuca	Rare	IV germ – microprop	Ronse (1990)
	Habenaria radiata	Rare	IV germ	nagayoshi *et al.*
(1996)	*Hetaeria cristata*	Prop for reintro and hort	IV germ, Rhozome nodes	Yam and Weatherhead (1990)
	Platanthera praeclara	Rare	IV germ	from and Read (1997)
	Renanthera inschootiana	over-coll, hab loss, prop for trade	Lf bases – adv buds	Seeni and Latha (1992)
	Spiranthes magnicamporum	Rare, prop for hort	IV germ	Anderson (1991)
	Spiranthes parksii	Less than 2000 in the wild	IV germ – microprop	Christenson (1988)
	Vanda coerulea	Over-collection	lf bases – adv sh	Seeni (1990)
	Vanilla walkeriae	Prop for preserv	Nodes	Agrawal *et al.* (1992)
Osmundaceae	*Todeea barbara*	Rare	Iv spore germ	oliphant (1989)
Pandanceae	*Sararanga sinuosa*	Rare, endemic	IV germ to microprop	Fay (1992)
Papaveraceae	*Meconopsis Paniculata*	Seed germ and survival low, hab loss.	Hypocots – callus – adv sh	Sulaiman (1994)
	Mecanopsis simplicifolia	Hab loss, seedling mortality	Seling explants – callus – adv sh	sulaiman and babu (1993)
Pinaceae	*Picea omorika*	Rare, grown for hort	IV germ – Sh – som emb. adv sh	Budhmir and Vujicic (1992)
Piperaceae	*Peperamia reticulata*	Rare, endemic	sh tips, nodes – microprop	krogstrup *et al.* (1990)
pitttosporaceae	*Pittosporum balfourii*	Rare, endemic	Sh tips, noeds – microprop	krogstrup *et al.* (1990)
Plunbaginaceae	*Limoniium calaminare, L.*	Rare, endemic, preserve germplansm	Iv germ –	Martin and Perez microprop
(1995)	*dufourei, L. gibertii*	for breeding		
	Limonium estevei	Few seeds available, Hort value	IV germ – microprop	Martin and Perez (1992)
	Limonium thiniense	Rare, endemic	Herb seeds, Iv germ – microprop	Lledo *et al.* (1996)
Podophyllaceae	*Podophyllum hexandrum*	Over-coll, med	Zyg emb – callus – som emb	Arumugam and bhojwani (1990)
Polygonaceae	*Rheum emodi*	over-coll, med	Sh tips	Lal and Ahuja (1993)
primulaceae	*Primula scotica*	Rare, endemic	IV germ – microprop	Benson *et al.* (1998)
Proteaceae	*Grevillea scapigera*	Rare	Sh tips, nodes – microprop, also	bunn and Dixon (1992)

Table 13.1 contd.

Family	species	Status	Propagation methods	References
			lf – adv sh	
Ranunculaceae	*Aconitum heterophyllum*	Over-coll, med	Sh tips – microprop Lf, petiole – callus – som emb	Giri *et al.* (1993)
	Delphinium malabaricum	Low seed set, seed dormancy	Infloresc nodes	Agrawal *et al.* (1991)
Restionaceae'	*Hopkinsia anoectocloea, Lepidobolus 'contorta', Loxocarya 'magna'*	Rare, low seed germ, slow prop	Embryo germ	Meney and Dixon (1995)
Rosaceae	*Cowania subintegra*	over-grazing, poor reproduction	sh tips – microprop	jakobek *et al.* (1993)
Rutaceae	*Citrus assamensis, C. indica, C. latipes*	Rare	Sh tip – microprop	baruah *et al*, (1996)
Rutaceae	*Citrus halinii*	Prop ofr germplasm preserve	IV germ, adv sh	Normah *et al.* (1997)
	Diplolaena andrewsii	Rare	Sh tip – microprop	Roseto *et al.* (1992)
	Drummondita ericoides	Rare	Sh tip – microprop	Rosetto *et al.* (1992)
	Phebalium equestre, P. hillebrandii	Rare, endemic, prop for hort, difficult to germ	sh tips, nodes – microprop	Jusaitis (1995)
Salicaceae	*Salix hallaisanensis*	Rare	Node – microprop	lee *et al.* (1995)
	Salix tarraconensis	Rare	Nodes – microprop	Amo-Marco and Lledo (1996)
Scrophulariaceae	*Isoplexis canariensis*	over-coll, med	IV germ – microprop	Arrebola *et al.* (1997)
	Penstemon haydenii	Rare	Nodes	Lindgren and Mccown (1992)
Picrorhiza kurroa	Rare, med		Sh tips, nodes – microprop	Lal *et al.* (1988); upadhyay *et al.* (1989)
Sterculiaceae	*Sterculia urens*	Over-coll	IV germ – microprop	Purohit and dave (1996)
	Trochetiopsis Erythroxylon, T. melanoxylon	Rare, endemic, low Germ rate	IV germ	Fay (1992)
Stylidiaceae	*Stylidium coroniforme*	Rare	Iv germ – microprop	McComb (1995)
Theaceae	*Camellia crapnelliana, C. granthaminana, C. hogkongensis*	Rare	Iv germ to microprop	Siu and Weatherhead (1995)
Thyrsopteridaceae	*Cibotium schiedei*	Rare	IV spore germ	Fay (1992)
Turneraceae	*Mathurina penduliflora*	Rare, endemic	Sh tips, nodes – microprop	krogstrup *et al.* (1990)
Valerianaceae	*Nardostachys jatamansi*	Over-coll, med root – adv sh	petiole – callus–	Mathur (1992)
	Valeriana wallichii	Rae, med	Sh tips, nodes – microprop	Mathur *et al*, (1988)
Zamiaceae	*Ceratozamia hildae, C. mexicana*	Rare	megagam – adv sh, Zyg emb – som emb	Chavez *et al.* (1992)

[1] Abbreviations: serm, germination; hab, habital; hort, horticultural value or value for cultivation; med, medicinal value; over-coll, over-collected; preserve, presevation; pop, population; prop, propagate; rare, endangered, threatened or rare; reintro, reintroduction; res, reseacrch; slow prop, slow to propagate.

[2] Procedures are given in a sequential format, with the steps in order, abbreviated and separated by hyphens. Abbreviation: adv, adventitious; cal, callus; fl, flower; imm, immature; infl, inflorescence; lf, leaf; microprop, micropropagation, using axillary ud ougrowith; som emb, somatic embryos; sh, shoot(s).

dangered species, and *in vitro* propagation techniques have been used successfully for a number of rare or endangered species. For specific details of the micropropagation techniques see Lynch.

Certain groups of plants have received particular attention. Techniques of *in vitro* seed germination have been applied to a number of rare orchid species, while endangered cacti and succulents and insectivorous plants and lilies are other groups which have been propagated *in vitro* by several labs. Laboratories in Australia, Spain, India, and Hawaii, among others, have focused on propagating their endemic floras, while laboratories in England, Denmark, Spain and elsewhere have also directed attention to propagating the endemic flora of islands such as St Helena, Gran Canaria, and Rodrigues, among others. Other programmes around the world have applied *in vitro* propagation techniques to a wide variety of native and exotic endangered species.

The variety of approaches used reflects the variety and flexibility of tissue culture techniques, described in more detail elsewhere in this volume. These approaches will be described briefly here, with some examples given in the specific context of endangered species conservation. They will be grouped into protocols which utilize seeds and those which begin with vegetative tissues.

In Vitro Propagation Using Seed

When seeds of endangered species are available, they are generally preferred for propagation, in order to maintain the maximum genetic diversity. Most endangered species produce seeds, but in some cases they are few in number or they may be difficult to germinate. When very few seeds are available, *in vitro* germination is often used to produce sterile seedlings, which are then used to provide shoot tips and nodes as explants for micropropagation. This approach has been used for a number of species. including *Gentiana lutea, Limonium* spp. and *Nepenthes khasiana*, as well as for many others.

When conventional procedures, such as stratification, fail to break seed dormancy or the rate of germination is very low, embryo culture may be useful. Some forms of dormancy are overcome by removing the seed coat, as with *Trochetiopsis* spp. from

St Helena and *Aster vialis*, a rare species from Oregon. In some cases, growth regulators have been added to stimulate germination. Alternatively, growth regulators may be used to stimulate direct somatic embryogenesis or shoot formation from the embryo tissue, as with *Primula maguirei*, a rare species from Utah, or to produce embryogenic callus, as with *Podophyllum hexandrum*.

In some cases, seeds have particular requirements for germination which are not easily met by conventional germination procedures. For example, *Pholisma sonorae,* and endangered parasitic plant of the south-western United States, requires the presence of host root tissues for germination, and it has not been possible to germinate the seeds *ex situ*. Other root parasites have been successfully germinated *in vitro*, and these techniques could be applie to seeds of *P. sonorae*.

Similarly, seeds of a number of rare orchid species have been asymbiotically germinated *in vitro*, such as *Vanilla alkeriae, Getaeria cristata,* and *Cypripedium reginae,* among others. In cases such as *Spiranthes magnicamporum,* symbiotic cultures of seeds and fungus have been established *in vitro,* stimulating germination further. Germinated orchid seeds have also been used to initiate cultures for micropropagation.

Propagation Without Seeds

For some endangered species, propagation by seed is not an option. Seed viability can be low, as with *Rauvolfia micrantha* and others, while in cases such as *Haworthia* spp., *Paronychia chartacea, Adhatoda beddomei*, and *Delphinium malabaricum* little or no seed is produced. With species such as *Artemisia granatensis* and *Limonium estevei*, so few plants were availble that taking seeds was restricted, while with species such as *Trillium persistens,* germination from seed is very slow. In the majority of cases, when seeds are not availble, propagation *in vitro* is accomplished by culturing shoot tips or nodes from field or greenhouse grown plants and stimulating the outgrowth of axillary shoots. This is preferred, since genetic changes appear to be less likely when preformed meristems are used for propagation.

The growth habit of some species is such that the culture of apical or vegetative lateral buds would irreversibly damage the plant. In the case of monopodial orchids, such as *Phalaenopsis,* the culture of dormant buds from inflorescence nodes has been used to over-come this problem. With *Delphinium malabaricum,* inflorescence nodes have also been used, since the single apical bud also grows at soil level, making it difficult to estabilsh uncontaminated cultures.

Although the culture of preformed meristems is generally preferred, because of their genetic stability, there are situations where buds are unavailable or difficult to culture or where more rapid propagation methods are desired. Organogenesis or embryogenesis has been obtained from vegetative tissues of *Meconopsis simplicifolia, Dionea muscipula, Agave victoria-reginae,* and *Haworthia* spp., among others. Species in the Liliaceae and Amaryllidaceae are often propagated using bulb scales or similar tissue, whereas protocorm-like bodies have been produced from leaf segments of monopodial orchids. In the case of *Nardostachys jatamansi,* which naturally forms buds from its roots, petiole callus was used to initiate adventitious roots *in vitro*, which were then stimulated to form buds.

GENETIC DIVERSITY AND GENETIC STABILITY

While tissue culture is a powerful tool for multiplying individuals of a particular genetic line, it is a clonal process and, at first, may appear to contradict the goal of preserving genetic diversity. However, genetic diversity is maintained by culturing each individual available for propagation as a unique and separate one.

A related concern is that of somaclonal variation, or the introduction of genetic changes into and otherwise clonal line. Plants obtained from preformed buds generally have a lower frequency of change than those from direct adventitious sources, while those from callus appear most likely to undergo changes. As discussed above, a number of factors are involved in developing a protocol for propagating an endangered species, and at times,buds cannot be used. In those cases, it may be necessary to regenerate adventitious shoots, but it may then be possible to propagate those shoots by axillary bud outgrowth. Another approach was taken with the micropropagation of the rare *Hackelia venusta* from the northwestern US. Cultures were grown and propagated on a minimal level of growth regulators, in order to minimize the potential for somaclonal variation. On the other hand, somaclonal variation has been suggested as a tool for increasing the genetic diversity in species with a vey narrow genetic base, such as the Easter island endemic, *Sophora toromiro*.

IN VITRO PROPAGATION OF ENDANGERED SPECIES

In vitro propagation of endangered plant species is generally undertaken to increase the numbers of individuals available of extremely rare and endangered species. Some species such as the Hawaiian *Cyanea pinnatifida*, are represented by only one individual in the wild. Small parts of tissue were removed from that plant and taken to the Lyon Arboretum for culture. There they were successfully propagated and reintroduced into the wild. Other species, such as *Sophora toromiro* are extinct in the wild, and the few plants which have been maintained *ex situ* havge been used for *in vitro* work. When population numbers are low, the number of *in vitro* plants can readily surpass the number of plants in the wild.

When species have been over-collected by hobbyists or for medicine, food, or fragrance, *in vitro* propagation can provide an alternate source of plants and alleviate pressures on wild populations. Certain orchids, cacti, and wildflowers as well as a number of medicinal species have heen propagated for this reason. On the other hand, some rare species may be relatively unknown, but could have horticultural or other value if enough material is made available for breeding and development.

Tissue culture can also be used when wild grown plants are difficult to propagate for *ex situ* preservation in botanical gardens. The Center for Plant Conservation coordinates botanical gardens in the united States to monitor and grow endangered species *ex situ*. When traditional propagation is difficult, *in vitro* techniques are used. Such plants can be used as a source of seed for long-term storage, and if seed is not produced, the tissue culture lines themselves can be cryopreserved. Propagated plants might also be used for *ex situ* studies on the biology of endangered plant species.

If wild populations become severely reduced or lost, propagated plants can also be used for reintroduction. This has been attempted with several species including *Artemisia granatensis, Nepenthes khasiana, Mammillaria san-angelensis, Senecio hadrosomus, Agave victoria-reginae, Bletia urbana* and several other orchid and rare Hawaiian species, including *Cyanea pinnatifida. Rubus humulifolius,* and endangered species of Finland, known from only ten plants, was propagated *in vitro* to yield 1500 plants and was replanted in a site near its original locality.

APPLICATION OF PRESERVATION TECHNOLOGIES

Traditional methods for preserving plant germplasm *ex situ* have included growing plants in botanical gardens and arboreta and banking the dried seeds and spores at refrigerator (4°C) or freezer (–18° or –20°C) temperatures. The development of cryopreservation, or storage in liquid nitrogen (at – 196°C), has provided a technology for even more stable, long-term storage of living tissues.

Several laboratories have applied cryopreservation protocols to the seeds of a variety of endangered species. A cooperative agreement between the National Seed Storage Lab of the US Department of Agriculture and the Center for plant Conservation was developed to store seeds of endangered US species at the NSSL facility. Cryopreservation is being applied to the seeds of the endangered and rare flora of Western Australia at Kings Park and Botanic Garden, while seeds of endangered and threatened species of Ohio are cryopreserved at the Cincinnati Zoo and Botanical Garden. Several other laboratories have also developed cryogenic storage facilities for seeds of native flora.

The majority of species currently stored in liquid nitrogen are those with orthodox, or desiccation tolerant seeds. When dried, orthodox seeds generally survive liquid nitrogen exposure with little or no damage. In some cases, however, seeds may be orthodox, but short-lived unless they are carefully dried and frozen either at –20°C or in liquid nitrogen. Two examples are the short-lived seeds of *Plantago cordata* and *Salix myricoides*, listed as endangered and potentially threatened in Ohio, respectively. These have been successfully dried, cryopreserved and banked in liquid nitrogen.

Other species have desiccation sensitive, or 'recalcitrant', seeds. These seeds cannot usually survive drying, and in the hydrated state they do not survive exposure to liquid nitrogen. However, cryopreservation is being increasingly applied to the excised embryos of recalcitrant seeds from tropical tree species and new approaches to recalcitrant seed cryopreservation are presently being considered. In general however, recalcitrant seeds do pose particular problems for long-term germplasm storage. Seeds of some large-seeded temperate trees, some wetland species and some climax species from the moist tropics fall into this category. Wetlands and moist tropical forests are two habitats that are particularly threatened, increasing the need for *ex situ* germplasm storage of species from these areas. Studies are underway at the National Seed Storage Laboratory to determine the extent of recalcitrance in seeds of endangered species from the rainforests of Hawaii, so that seed storage protocols can be developed for these species. A similar study is being conducted on endangered species from Ohio wetlands by the Cincinnati Zoo and Botanical Garden. In both cases, it appears that the majority of the species under study are not recalcitrant.

Cryopreservation of 'non-seed' tissues, such as immature embryos or *in vitro* cultures offers an alternative approach to the used for the preservation of recalcitrant species. These procedures centre around the techniques of slow freezing, vitrification, and encapsulation-dehydration. Tissues most commonly used for cryogenic storage are: shoot tips from *in vitro* cultures, excised zygotic embryos and embryonic axes, somatic embryos and embryogenic or organogenic cell or callus lines.

When endangered species are propagated by tissue culture, the tissue culture lines can be cryopreserved. Cyropreservation of *in vitro* tissue from endangered species has been accomplished using a slow freezing protocol with shoot tips of *Grevillea scapigera* and organogenic callus of *Dioscorea caucasica* and *D. balcanica*. Other species have been cryopreserved using encapsulation dehydration, including *Centaurium rigualii,* endemic to the Iberian peninsula, *Aster vialis* and *Sisyrinchium sarmentosum*, two endangered species from Oregon, and *Cosmos atrosanguinensis,* which is cultivated but is extinct in the wild.

In addition to tissues of seed plants, spores and gametophytes of pteridophytes and bryophytes can also be cryopreserved. Non-chlorophyllous fern spores are generally desiccation tolerant and adapt well to liquid nitrogen (LN) storage, although some are short-lived, such as those of the endangered tree fern, *Cyathea spinulosa.* These were dried and exposed to LN, with over 93 per cent recovery. Chlorophyllous spores of at least some ferns, although they are generally short-lived, also can by dried and cryopreserved or cryopreserved using the encapsulation-dehydration technique. Moss spores of several species have also survived cryopreservation.

Some gametophytes of mosses and liverworts are naturally desiccation tolerant and when this is the case, they can be air dried and frozen directly in liquid nitrogen. When gametophytes are sensitive to drying, pre-culture with abscisic acid (ABA) is sufficient to induce desiccation tolerance in some species of bryophytes. In other cases the encapsulation-dehydration technique has been useful in preserving both tropical and temperate brytophytes and fern gametophytes through desiccation and subsequent liquid nitrogen exposure. Slow freezing has also been used to cryopreserve protoplasts of *Marchantia*, while pre-culture with mannitol or ABA and proline has been shown to provide protection of moss tissues through slow freezing protocols. These techniques should be readily transferable to rare or endangered bryophyte and pteridophyte species.

Slow growth is another option for preserving *in vitro* cultures of endangered species for rmedium-term storage of several months to several years. Good survival has been reported for shoots of *Centaurium rigualii* ror three years, *Picrorhiza kurroa* for ten months, *Saussurea lappa* for 12 months, and of *Coronopus navasii, Lavatera oblongifolia,* and *Centauriuim rigualii* for six months; when stored at 5°C. Similarly, *Drosera* spp. and *Dionaea muscipula* have been maintained for up to ten months at 0 – 6°C. Rauvolfia spentina remained healthy after 15 months at 15°C, although lower temperatures were deleterious. A number of rare Hawaiian species are maintained at the Lyon Arboretum on minimal medium *in vitro*, and only transferred at six month intervals.

COLLECTION BIOTECHNOLOGY

In vitro collection, or IVC, is the initiation of tissue cultures in the field. it can be used to broaden germplasm collection to include species for which seeds are not available and for which cuttings may be difficult to maintain or transport.

IVC is a very flexible technique and can be adapted to a variety of situations. Either partial or full sterilization of the tissue is made on site and the tissue is transferred to containers of medium for transport back to the lab. In some cases, such as with *Cocos nucifera*, tissues have been collected with minimal treatments in the field. Once they were transported back to the lab, they were resterilized and dissected further for culture. In other cases, sterilization and dissection have been completed in the field and the growth and development of the cultures initiated at that point. IVC has been used to collect a variety of plant tissues, including orchid seeds, embrya apical or nodal buds, and leaf and stem tissue.

Different strategies have been usd to minimize contamination in IVC cultures. In some cases, a portable glove box was used to reduce contamination from ambient sources, whereas in other cases the work has been done quickly in the open air. Internal contamination can be a more serious problem than ambient contamination, since many plants harbour endophytic fungi and bacteria. However, the use of fungicides and antibiotics in the medium can reduce this contamination to a workable level.

The initiation of *in vitro* cultures in the field can facilitate the transport of the tissues. Because of the small size of the explants, more material can be transported, compared to the transport of whole plants or cuttings. In addition, the cultures are initiated with fresh material which can begin the process of growth *in vitro* immediately, compared with whole plant materials which may undergo some deterioration in transport before they are planted *ex situ*. Finally, the transport of clean plant materials *in vitro* generally facilitates their movement through international border inspections.

IVC could have broad applicability in collecting materials from endangered plant species when seeds or spores are not available or to supplement seed collection, when seeds are of low viability or are difficult to germinate. Wild callected seeds of *Drosera rotundifalia* a potentially threatened species in the state of ohio, proved difficult to germinate in early studies. Thus, as a supplement to seed collection, leaf tissues were collected by IVC and plants were successfully regenerated from these. Seed collection is prohibited from several other Ohio species, because of their rarity. IVC will be used as an alternative, to collect vegetative tissue for propagation and preservation, with minimal disturbance to the plants in the field.

IVC can be used as a source of material for both the propagation and preservation of endangered plant germplasm. For example, leaf and bud tissue collected by IVC from *Brunfelsia densifolia*, a rare Puerto Rican tree growing at the Fairchild Tropical Garden in Florida, was transported to the Cincinnati Zoo and Botanical Garden where plants were regenerated from the cultures and tissue was cryopreserved.

The limitations of IVC are those of tissue culture in general, since conditions for growing some species *in vitro* have not yet been developed. However, the number and variety of species which have been successfully propagated *in vitro* continues to grow, and this will, in turn, be reflected in the widening applicability of IVC techniques for the collection of rare or endangered plant germplasm.

MOLECULAR TECHNIQUES OF PLANT CONSERVATION

The development of molecular techniques has opened the door to a number of areas relevant to monitoring the stability and diversity of endangered plant germplasm held *in situ* and *ex situ*.

Molecular techniques, particularly RAPD analyses, are being used to monitor the genetic diversity of populations of rare species, as well as to define species themselves, all of which impact on endangered species management. A high level of diversity was found for the ten known populations of *Banksia cuneata,* an endangered species of south western Australia. RAPD analysis also indicated that the one known individual of *Eucalyptus graniticola* was a hybrid of two more common species, rather than an endangered relict species. Thus, rather than reinforcing the population *in situ*, the species was backed up *ex situ*. At the Chicago Botanic Garden, RAPDs are being usd to compare populations of the rare *Cirsium pitcheri*. DNA from Indiana and Wisconsin populations is being analysed and compared with DNA from 130-year-lod herbarium specimens of the extirpated Illinois population.

Methods for detecting somaclonal variation in tissue culture propagated species have traditionally included phenotypic evaluation, isozyme analysis, chromosome counts, and DNA cytofluorometry. For example, isozyme analysis and chromosome counts were made to evaluate plants micropropagated from *Centaurium rigualii,* an endangered species of Spain, and no variations were found using these methods. The use of RAPSs can provide an even more precise tool for the detection of somaclonal variation in micropropagated endangered plants.

In preserving plant germplasm, DNA can be banked as a backup or supplement to the storage of living tissue, or may be used when living tissues cannot be stored. Techniques for isolating DNA from dried tissue, as well as dried frozen material have been developed, making collection and banking from wild, remote species a possibility. Libraries of DNA from rare or endangered species are being set up to store this information for future use and recommendations and guidelines for DNA banking have been made.

Molecular technologies offer the possibility of collecting ancient DNA from extinct species, as well. Within the past decade, the isolation of DNA fragments from dried herbarium specimens as well as from fossil materials has been demonstrated. A fragment of DNA, identified as part of the RuBisCo gene, was obtained from a fossil leaf compression of the extinct *Magnolia latahensis*. This was compared with gene sequences from extant Magnoliaceae and other species in order to determine the genetic relationship of the extinct and extant species. A portion of this same gene was also obtained from fossil *Taxodium*.

Spain is the European country with the richest and most exclusive flora. When the Canary Islands are included, there are approximately 1300 endemic plant species of which many have a high biological interest. Unfortunately a high number of these species are on the verge of extinction. Recent surveys of these endangered plant species of continental Spain and the Balearic and Canary Islands denote around 600 endemic species which, according to the categories followed by IUCN, are threatened to some degree. The current situation could develop, in the near future, not only with the loss of very important germplasm useful to breeding programmes, but also with reduced biodiversity which would undermine ecosystem stability.

Although some of these threatended species are protected by local or international laws, measures for the conservation of endangered plant species in their natural habitats are only very slowly being introduced. Therefore, it is highly desirable to develop appropiate techniques for *ex situ* conservation in order to enable a quick and efficient response to the most critical situations.

There are two different strategies in which *in vitro* culture techniques can be used in the conservation of endangered species. The first one is the use of miocropropagation techniques in order to rapidly increase the number of individuals in species with reproductive problems and/or with extremely reduced populations. Plant material thus obtained can be of great value for research, living collections, to reduce pressure of botanical collection on the natural population and, if considered appropriate, for plant introduction programmes. The second strategy is the development of *in vitro* storage techniques which are particularly useful when the conservation of seeds is not possible. These techniques allow *ex situ* conservation at a vey low maintenance cost.

Using these techniques, a large amount of plant material can be produced in a short period of time from a minimum of starting plant material, and hence, with minimal impact on the native population. Another advantage of *in vitro* culture techniques is that they can be an alternative to seed banks with highly restricted species, where the simple act of gathering seeds in the natural population may affect its survival. However, the convenience of these techniques is controversial, particularly with regard to the possible genetic variability produced throughout the process. In these circumstances, monitoring regenerated plantlets through somaclonal variation detection techniques may be of interest to eliminate undesirable variants.

The choice of an adequate explant, the micropropagation cycle and the type and concentration of growth regulators may all contribute to reduce the appearance of somaclonal variants, but it will not completely eliminate the probability of inducing them. On the other hand, the limted availability of starting plant material especially prevalent with threatened plant species, further complicates the process of micropropagation, leaving little margin for the previous considerations. In this chapter we will describe several studies on the preservation of Spanish endemic plant species using biotechnology.

MICROPROPAGATION OF ENDANGERED PLANTS

In the past years there has been intense activity with regard to the development of micropropagation protocols for Spanish endemics. includes some of the Sapnish endangered endemics on which micropropagation methods have been published. Actually, botanical gardens, universities and other research institutions have worked on a much longer list of Spanish endangerd endemics, but the results have not always been published. Information on much of this unpublished activity has been gathered by Iriondo *et al.* (1994)

It is remarkable that 39 per cent of the species listed in belong to the genus *Limonium*. More than half of these species (56 per cent) belong to two families: *Plumbaginaceae* and *Asteraceae*. These high percentages are easily understood when it is considered that nearly half of all the Iberian endemics belong to just 27 genera, one of which is the genus *Limonium*. In the Iberian Peninsula there are 86 endemic, mostly endangered or rare, species of *Limonium*. In contrast, only 8 per cent of the species are monocotyledons.

TABLE 13.2 ENDANGERED ENDEMICS ON WHICH MICROPROPAGATION METHODS HAVE BEEN DEVELOPED FOR CONSERVATION PURPOSES

Family	Species	Reference
Amaryllidaceae	*Narcissus longispathus* Pugsley	Clemente (1991)
	Narcissus nevadensis Pugsley	Clemente (1991)
	Narcissus tortifolius Fdez. Casas	Clemente (1991)
Asteraceae	*Artemisia granatensis* Boiss.	Clemente *et al.* (1991)
	Atractylis arbuscula Svent, & Mich.	Gonzalez *et al.* (1989)
	Centaurea balearica Rodr.	Estades and Medrano (1990)
	Centaurea carratracensis lange	Clemente (1991)
	Pericallis hadrosoma (Svent.) B. Nord.	Ortega and Gonzalez (1990)
	Senecio hermosae Pitard	ortega and Gonzalez (1985)
Brassicaceae	*Coronopus navasii* Pau	Iriondo and Perez (1990)
	Vella lucentina M.B. Crespo	Lledo *et al.* (1993)
Cistaceae	*Cistus heterophyllus* Desf. suibsp. *carthaginensis* (Pau) M.B. Crespo & Mateo	Arregui *et al.* (1995)
	Helianthemum polygonoides Pein., Mtnez-Parras, Alc. & Esp.	Iriondo *et al.* (1985)
Euphorbiaceae	*Euphorbia handiensis* Burchd.	Gonzalez *et al.* (1988)
Gentianaceae	*Centaurium rigualii* Esteve	iriondo and Perez (1996a)
Globulariaceae	*Globularia ascanii* Bramwell & Kunkel	bramwell (1990); Cabrera (1995)
Lamiaceae	*Thymus richardii* Pers.	Eastades and Medrano (1990)
Leguminosae	*Lotus berthelelotii* Masferre	ortega (1982)
	Lotus kundelii (Esteve) Bramwell & Davis	Bramwell (1990)
Malvaceae	*Lavatera oblongifolia* Boiss.	Irionds and Perez (1992)
Plumbaginaceae	*limonium arborescens* (Brouss.) O. kuntze	Fay (1993)
	Limonium calaninare Pignatti	martin and Perez (1995)
	Limonium dufourii (Girard) O. Kuntze	Lledo *et at.* (1993) Martin and Perez (1995)

Table 13.2 Contd.

Family	Species	Reference
	Limonium estevei Fdez. Casas	Martin and Perez (1992)
	Limonium fruticans (Webb) O. Kuntze	Fay (1993)
	Limonium gibertii (Sennen) Sennen	Martin and Perez (1995)
	Limonium imbricatum (Webb & Berthel.) Hubbard	Fay (1993)
	Limonium parvibracteatum Pignatti	Lledo *et al.* (1993)
	Limonium redivivum (Svent.) Kunkel & Sund Sund.	Fay (1993)
	Limonium spectabile (Svent) Kunkel & Sund	Fay (1993)
	Limonium rigualii M.B. Crespo & Erben	Lledo *et al.* (1993)
	Limonium santapolense Erben	Lledo *et al.* (1993)
	Limonium thiniense Erben	Lledo *et al.* (1996)
Scrophulariaceae	*Antirrhinum microphyllum* Roth.	Gonzalez-benito *et al.* (1996)
Thymeleaceae	*Daphne rodriguezii* Teixidoe	Estades and Medrano (1990)

Research on *in vitro* storage of Spanish endemics is much more scarce. Minimum growth techniques have been used with *Centaurium rigualii, Coronopus navasii, Lavatera oblongifolia, Limonium calaminare, L. catalaunicum, L. dichotomum, L. dufourii, L. estevei* and *L. gibertii* with relative success. In general, reduction of temperature was the most effective way of decreasing growth. Cultures stored at 5°C in MS medium alone or supplimented with 4.44 μM BAP + 0.54 μM NAA for periods ranging from four to six months maintained a good survival rate.

In some cases, as in the *in vitro* storage protocol for *L. estevei,* the maintenance of the cultures in darkness together with the reduction of temperature (5°C) allowed for a longer period of conservation with good recovery.

On the other hand, in *Coronopus navasii* a higher survival rate was achieved when cultures were stored uner a 16 hour light photoperiod. Cryopreservation techniques have also been used in the conservation of *in vitro* propagated explants of Spanish endemics, and they will be discussed in the next section.

These techniques constitute an important tool for the conservation of the rich endemic flora of Spain, where an effective *in situ* conservation policy is not always possible for all species. At present, a good representation of endangered plants is being obtained through micropropagation. These plants are being used for scientific studies and for display in botanical gardens.

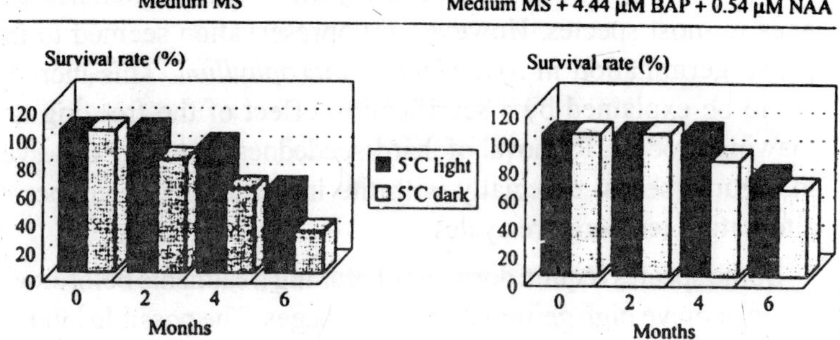

Figure 13.1 : *Survival rate after different periods of shoots of* Coronopus navasii *stored in* vitro *in different incubation conditions.*

CRYOPRESERVATION

Seeds of many Spanish endemics have been stored dry at –20°C in the Seed Bank run by the Department of Plant Biology at the Universidad Politecnica de Madrid. There is, however, an increasing interest in cryopreservation as some physiological or even genetic damage can occur in seeds stored for long periods under such conditions. Such deterioration could be of great concern when the species are somehow threatened due to the scarcity of their populations, their low seed production or their habitat degradation. The development of cryopreservation protocols of vegetative explants is also of interest for those species whose micropropagation has already been carried out.

Cryopreservation of Orthodox Seeds

Studies largely concern species of the families Asteraceae, Betulaceae, Brassicaceae, Cariophylaceae, Cistaceae, Leguminosae, Plumbaginaceae and Scrophulariaceae. Cryopreservation was carried out by direct immersion of seeds in liquid nitrogen, where they were maintained for one or 30 days. Thawing took place at room temperature.

Most seeds studied either had a low initial moisture content or this was reduced by desiccation with silica gel (3.0 to 8.4 per cent fresh weight basis). Only seeds of *Gypsophila struthium* had a slighty higher moisture content (10.8 per cent). Desiccation did not improve seed survival after storage in liquid nitrogen in those species where this factor was studied. Storage for one or 30 days in liquid nitrogen did not significantly affect germination percentages in most species. However, cryopreservation seemed to improve germination in *Antirrhinum microphyllum.* This increase could be explained by a scarification effect of the freezing and thawing process. Removal of hard-seededness has been observed in legume seeds, especially with the increase of the number of alternating temperature cycles.

Some species require dormancy breaking treatments before sowing to achieve high germination percentages. The possible interaction of this treatment with cryopreservation has been studied in three Cistaceae and on Asteraceae species. Seeds of *Cistus osbeckiifolius, Helianthemum polygonoides* and *H. squamatum* were placed in boiling water and soaked for up to 24 hours as the water cooled down to room temperature. *Onopordum nogalesii* seeds were immersed in a solution of giberellic acid (1000 $mg\,l^{-1}$) for 24 hours at 20°C. For cryopreserved seeds, dormancy breaking treatments were performed after thawing. The results indicated that the dormancy breaking treatment applied increased germination in control and frozen seeds in the first three species. In *C. osbeckiifolius* and *H. polygonoides* a significant cryopreservation × dormancy-breaking interaction was found ($p < 0.05$). However, cryopreserved and dormancy treated seeds never showed lower germination than the dormancy treated non-cryopreserved seeds.

Other factors studied include thawing speed and humidification of seeds before sowing. Seeds of two endemic species (*Centaurea hyssopifolia* and *Limonium dichotomum*) were cryopreserved by immersion in liquid nitrogen for one or 30 days. Seeds were then thawed either at room temperature (slow thawing) or in a 40°C water bath (fast thawing). Half of the seeds were subjected to humidification before sowing by placing them in a sealed container above water at room temperature for three days. The results showed that for both species there was not a significant effect of cryopreservation or the speed of thawing on germination .

TABLE 13.3: RESULTS OF SEED CRYOPRESERVATION

Family	Species	MC[a]	GT	Germination (%) C[b]	LN (1 day)[c]	LN (30 days)[c]	Reference
Asteraceae	*Onopordum*	10.7	C	9[d]	6[d]	8[d]	Iriondo *et al.* (1992)
	nervosum Boiss	6.5	B	72	75	87	Gonzalez-benito *et al.* (1998a)
		5.0	C	9[d]	3[d]	12[d]	Iriondo *et al.* (1992)
	O. nogalesii Svent	7.02	B	6[e]	1[e]	7[e]	Gonzlez-Benito *et al.* (1998b)
		7.0	B	4	4	3	Gonzalez-benito *et al.* (1998b)
	Centaurea	6.8	E	65[d]	73[d]	67[d]	Gonzalez-Benito
	hyssopifolia Vahl	6.8	E		64[df]	62[df]	*et al.* (1998a)
		6.8	E	48	47	38	Gonzalez-benito
		6.8	E		52[f]	52[f]	*et al.* (1998a)
Betulaceae	*Betula celiberica*	10.6	C	3[d]	4[d]	3[d]	Iriondo *et al.* (1992)
	Rothm. et Vase.	6.1	C	1[d]	3[d]	7[d]	Iriondo *et al.* (1992)
Brassicaceae	*Coronopus navasii* (Cav.) DC.	ND	D	30	34	40	Gonzalez-benito *et al.* (1998b)
	Diplotaxis virgata (Cav.) DC.	7.7	D	90	93	88	Gonzalez-benito *et al.* (1998b)
	Iberis pectinata Boiss	7.2	A	97	97	98	Gonzalez-Benito *et al.* (1998b)
	Vella paeudocytisus L.	7.3	d	97	98	100	Gonzalez-Benito *et al.* (1998b)
Cariophylaceae	*Gypsohila struthium* Loeft.	10.8	B	100	93	96	Gonzalez-Benito *et al.* (1998b)
Cistaceae	*Cistus osbeckiifolius* Webb ex Christ.	4.3	A	57[e]	74[e]	56[e]	Gonzalez-Benito *et al.* (1998b)
		4.3	A	30	25	24	Gonzlez-Benito *et al.* (1998b)
	Halimium	6.4	C	6[d]	1[d]	1[d]	iroindo *et al,* (1992)
	Artiplicifolium (Lam.) Spach	5.3	C	4[d]	2[d]	0[d]	iroindo *et al,* (1992)
	Helianthemum polygnoides	3.9	D	42[d]	42[d]	55[d]	Gonzlez-Benito *et al.* (1998b)
	Peinado *et al.*	3.9	D	7	6	4	Gonzlez-Benito *et al.* (1998b)
	H. squanatum (L) Dum. Cours.	ND	D	32[e]	29[e]	39[e]	Gonzlez-Benito *et al.* (1998b)
		ND	D	21	18	17	Gonzlez-Benito *et al.* (1998b)
Leguminosae	*C. atlantica*	11.0	C	4[d]	2[d]	0[d]	Iriondo *et al.* (1992)
	Browicz	7.8	C	2[d]	1[d]	0[d]	Iriondo *et al.* (1992)
	Onobrychis eriophora Desv.	7.7	C	48[de]	40[de]	–	Gonzlez-Benito *et al.* (1994)
		5.3	C	60[de]	70[de]	–	Gonzlez-Benito *et al.* (1994)
		6.0	C	42[d]	51[d]	54[d]	Iriondo *et al.* (1992)

Table 12.3 Contd.

Family	Species	MC[a]	GT	Germination (%) C[b]	LN (1 day)[c]	LN (30 days)[c]	Reference
		5.2	C	55[d]	46[d]	31[d]	Iriondo *et al.* (1992)
	O. peduncularis (Cav.) D.C. ssp *matritensis* (Boiss. et Reuter)Maire	ND	A	83	78	73	Gonzlez-Benito *et al.* (1998b)
plumbaginaceae	*Limonium dichotomum* (Cav.) Kuntze	8.4	E	83	82	83	Gonzlez-Benito *et al.* (1998b)
		8.4	E		84[f]	83[f]	Gonzlez-Benito *et al.* (1998b)
Scrophularia- -ceae	*Antirrhinum majus* L. ssp. *barrelieri* Boreau	ND	A	93	92	91	Gonzlez-Benito *et al.* (1998b)
	A. microphyllum Rothm	ND	A	38	52	55	Gonzlez-Benito *et al.* (1998b)

[a] MC, moisture content (fresh weight basis).
[b] C, control (unfrozen) seeds.
[c] Storage for 1 or 30 days in liquid nitrogen.
[d] Hunidification by placing seeds in a saturated atmosphere.
[e] Dormancy breaking treatment carried out.
[f] Fast thawing (water bath at 40°C).
ND, not determined.
GT, germination temperature: A, 15°C; B, 20°C; C, 25°C; D, 25°C (day) 16 h/15°C (night) 8 h; E, 10 days at 15°C, then 25/15°C

Humidification increased germination in *C. hyssopifolia* and did not have any effeect in *L. dichotomum*. This could have been due to the fact that the first species has seeds with hard sclerenchyma layers. No interaction between humidification and freezing in liquid nitrogen was observed in these two species.

Cryopreservation of Vegetatively Propagated Germplasm

Cryopreservation of nodal explants of *Centaurium rigualii* Esteve (Gentianaceae), an endemic species from the south-east Iberian Peninsula, has been studied using two different techniques: vitrification and encapsulation- dehydration. In the first study Sakai *et al.* (1990) vitrification solution was used. No survival was achieved unless explants had been previously cultured on semi solid medium containing 2 per cent (v/v) dimethyl sulphoxide or glycerol for two days, when 15 per cent survival was observed.

A protocol to improve survival of *C. rigualii* nodal explants after direct immersion in liquid nitrogen was developed using the encapsulation– dehydration technique to protect against ice crystal formation . Each node was included in an alginate bead, cultured in liquid medium with 0.75 M sucrose, desiccated with silica gel to around 20 per cent moisture content and subjected to immersion in liquid nitrogen. The preculture of nodes in 0.1 M sucrose (standard concentration) for three days improved survival after freezing and fast thawing (32 per cent) compared to non-precultured nodes (5 per cent). When node preculture was carried out in a higher sucrose concentration (0.3 M) for one day, survival after freezing was increased to 70 per cent on the eighth week of culture.

This technique has also been employed with nodal segments of another Spanish endemic species: *Antirrhinum microphyllum*. Different desiccation times and preculture media were studied. A survival rate of 85 per cent was obtained after cryopreservation when beads had been desiccated for four or five hours (21 and 20 per cent moisture content, respectively). However, the number of frozen nodal explants that showed bud elongation as a response after four hours desiccation was higher when the preculture had been carried out with 0.3 M sucrose (70 per cent vs 53 per cent), although this difference was not found after five hours of desiccation (68 per cent).

Preculture of explants on medium with a high sucrose concentration to improve survival after freezing has been reported in many encapsulation– dehydration protocols. In some cases no survival was observed after immersion in liquid nitrogen without preculture on 0.3 M sucrose. Similarly, survival was improved for cryopreserved nodal segments of *Dianthus hybridus* after a two day preculture on 0.6 M sucrose.

STABILITY ASSESSMENTS

The use of *in vitro* culture techniques bears the risk of somaclonal variation occurring. It is therefore necessary to assess the genetic stability of the cultures, particularly when the aim of the *in vitro* techniques is genetic conservation. Although a great number of spanish endemic species have been micropropagated, the occurrence of genetic changes has only been assesed in a few of them. Several techniques can be used to detect morphological variations and genetic changes. The latter can include studies on chromosome structure and number, or the use of molecular markers (isozymes or DNA markers). A combination of several techniques is recommended for the evaluation of the regenerated plants, especially when one of them is based on morphological characters.

In *Lavatera oblohgifolia*, shoots regenerated from calli showed relevant morphological variants. In some cases, shoot looked extremely thick and were surrounded by a large number of leaves arranged in an abnormal phyllotaxis. However, most of the regenerated plantlets that lived through the acclimatization process did not present markedly abnormal morphological features. Leaf morphology of regenerated *Lavatera* plantlets was further examined through computer-assisted image analyis techniques. No single variants were found in leaf shape, and no signification differences were found in petiole length among non-acclimatized vitroplants, acclimatized vitroplants and plants obtained from cuttings, a situation commonly described in plants of other species regenerated from calli, e.g. *Saintpaulia ionantha*. However, qualitative differences in leaf hairiness between non-acclimatized plants and the rest of the groups were found. Stellate hairs of non-acclimatized plants were shorter in height, narrower, and more transparent. No differences among regenerants were detected when stability was assessed with the following six isozyme systems: malate dehydrogenase (MDH), glutamate oxaloacetate transaminase (GOT), cytochrome oxidase (Cy-O),acid phosphatase (Ac-P), catechol oxidase (CO) and peroxidase (PER). Nevertheless, four different types of esterase (EST) zymograms were obtained from leaves of acclimatized regenerants derived from a single clone. The use of the esterase isozyme system has been discussed for this type of study as it is highly varible among tissues and artefacts easily appear.

L. oblongifolia has a chromosome number of 2n = 42 (x = 7, hexaploid) (Luque and Devesa, 1986). No changes in the ploidy level were found among regenerants, although chromosome counts with 2n = 36–38 suggest possible hypo-aneuploids.

In *Coronopus navasii,* neither morphological nor isozyme variation was detected among regenerants from a single clone. However, a chromosome duplication was detected in one of the acclimatized plants (2n = 64 vs 2n = 32) (Iriondo and Perez, 1991b).

Seven isozyme systems (PER, EST, GOT, Ac-P, MDH Cy-O and CO) were studied in *Centaurium rigualii* in the ninth subculture of a clonal line of shoot cultures. All isozyme systems but esterase showed recurring banding patterns. In this system, one of the somaclones diverged from the rest of the sample. However, when the somaclones were resampled at the twelfth subculture no differences were found in their esterase banding patterns. In these plants, caryological observations showed no departures from the natural chromosome number (2n = 20).

Molecular markers based on DNA offer the advantage of revealing changes occuring directly in DNA sequences. Some of these techniques, such as RAPD (random amplified polymorphic DNA) have the additional advantage of requiring a small quantity of DNA, which is an important factor when working with threatened species. Another benefit of this method is that no previous knowledge of the genome studied is required.

This method was emmloyed to evaluate the possible somaclonal variantion occurring i plants of *Limonium estevei* obtained through micropropagation. Six plants from each of five clonal lines subcultured every month for two years were tested with three different primers. The band patterns revealed differences between clonal lines, but intraclonal variation was not detected. The same technique was used in the assessement of genetic stability of plants of *Centaurium rigualii* which were maintained in *in vitro* conditions for six years with monthly subcultures. All the plants came from the same original clonal line. In this case, markers from 13 different primers were considered to establish the differences of the samples through an UPGMA (unweighted pair group method arithmetic average) analysis. From 31 plants tested, 15 different patterns were detected.

These examples show that somaclonal variation can occur and thus justify the need to monitor the stability of *in vitro* cultured plants at different stages of the process, especially when genetic conservation is the aim of these techniques.

PLANT DIVERSITY ASSESSMENT

Many researchers consider that the knowledge of the species genetic composition is essential for any comprehensive conservation plan. The study of the genetic composition and, therefore, the species diversity is important to interpret its demography, reproductive biology and evolution.

The knowledge of the population genetic variability is especially necessary when the conservation strategy is focused on *ex situ* conservation of plant material, which should be representative of the genetic variation in the natural population. This is even more relevant if introduction or reintroduction measures are to be considered.

There are only a few studies about the genetic composition of wild species, and the number is even lower when considering endangered or rare species. In recent years this kind of work has proliferated with the employment of molecular markers, firstly with the use of isozymes, and more recently with DNA markers (RFLP, DNA-fingerprinting, RAPD).

As previously mentioned, one of the main advantages that the RAPD technique offers is that it only requires a small amount of DNA and, therefore, a small amount of plant material, which is in many cases a limiting factor when working with endangered plants.

In the species in which the establishment of *in vitro* cultures is the chosen conservation strategy, assessment of the genetic diversity of the initial plant material guarantees the representation of the diversity of the natural population in the sample. At the same time, it can be used as a reference to check the occurrence of somaclonal variation after the *in vitro* culture period. This kind of strategy has been used in the conservation project for *Antirrhinum microphyllum* Undertakan in our laboratory. Using RAPD markers the genetic study of the natural populations and of the micropropagated plants is studied to check whether the material propagated is representative of the original population and its stability after culture.

Some other studies on the genetic diversity of natural populations of endemic Spanish plants using RAPD markers have been published: *Erodium paularense, Limonium cavanillesii, Limonium dufourii.* The information obtained from these works may be very useful for further studies or for the application of conservation strategies.

CONCLUSIONS

Throughout this chapter, we have shown some of the research where biotechnology has been applied in the conservation of Spanish endemics. It sholud be borne in mind that this type of work should be integrated in a broader perspective and not be an aim in itself. The final goal should be the species conservation. These studies are now being extended in some of the species already mentioned (e.g. *Erodium paularense* and *Antirrhinum microphyllum*), comprising demography population models, reproductive biology, etc. All this knowledge will allow the appropriate application of biotechnology to species preservation.

The two main questions to be considered in *ex situ* conservation pan of an endangered plant species are: 'What to conserve?' and 'How to conserve?'.

The selection of the species to preserve is based in many cases on economical factors. The priority of some species above others will often take into account the imminence and type of threat. In addition, what to conserve within a species has not always been so carefully considered or has not been based on scientific knowledge. It is precisely in this situation where molecular techniques can play an important role. Firstly, these methods can show the genetic diversity of the species populations. Secondly, they can be used to check whether the 'conserved population' represents the diversity of the original population. With this knowledge some other practical problems can also be solved, such as whether to preserve populations separately or in a single sample.

Economic factors also dictate the procedure of conservation. Seed banking is probably the easiest and cheapest way of preserving a wide range of genotypes of a certain species. When this option is not possible (e.g. as is the case for vegetatively propagated species and recalcitrant seeds), one should compare conventional methods (field collections) with using biotechnological approaches. If the species has recalcitrant seeds (i.e. they cannot be desiccated without viability loss), cryopreservation of embryos could be employed (Pence, 1995). In this or other cases, conservation of vegetative plant material could be considered through *in vitro* conservation or cryopreservation. The most appropriate technique for the conservation of a species can be selected after a careful analysis of the nature of the propagules, the techniques that could be used and the resources available (both material and human).

When *in vitro* conservation techniques are used, or when cryopreservation implies the use of *in vitro* culture at some point, special care should be taken to check the possible appearance of somaclonal variation. The genetic composition of the stored sample should be compared after different conservation periods with that of the sample at the starting point. It should also be considered that not only could genetic changes occur during this process, but that there also could be an accidental selection towards genotypes better fit for the *in vitro* technique employed. However, in some extreme cases where the number of individuals is very low, micropropagation is clearly justified. The emergency and seriousness of the situation weighed against the probability of somaclonal variation could make the latter irrelevant.

The final question to be answered could be: What for? The preserved sample could be used for research work or to display in botanical gardens. The convenience of the use of micropropagation in plant reintroduction programmes should be carefully studied. A great number of genetically cloned specimens obtained through this technique from a few individuals of the native population may not be of great value. If planted alongside the original population genetic impoverishment could possibly occur. On the other hand, if planted apart from the original population, its narrow genetic base would decrease the chances of adaptation to the habitat and survival in extremely mutable conditions. Studies on the genetic diversity of a given population, knowledge of the physiology of reproduction of the species and the estimation of the number of clone lines needed to adequately represent the native population's genetic diversity are all or great value when preserving endangered species.

We believe that further research is needed in order to determine correctly the value of these techniques was a method to avoid the extinction of threatened species. The experience obtained from the application of these techniques to different species, each one with its own characteristics, will be of great significance in this respect.

14

Algal Conservation

Algae are an ancient and extremely diverse group of plantlike organisms, with representatives of the blue-green algae (cyanobacteria) being present for the last 3550 million years . They range in morphology and size from microscopic picoplanktonic cyanobacteria (<2 μm in diameter) which are prokaryotic and closely resemble other eubacteria, a vairety of unicellular, multi-cellular, filamentous and thalloid forms, to giant kelps that may be upto 60 m long. Their taxonomy is problematic, but it is clear on the basis of both traditional taxonomy and modern molecular techniques that they are polyphyletic. The 'amount' of algal biodiversity, an in other groups of organisms, is largely unknown, however, the advent of molecular biological techniques and improvements in electron microscopy have greatly increased our knowledge-base and assisted in elucidating inter-relationships. Approximately 37000 species of algae have been recognized/described (Table) but estimates of the total number of algal species vary from a relatively conservative 40 000 to > 10 000 000.

From an economic perspective, algae are often considered a nuisance. Those associated with the water industry may think of them as a source of irritation—blocking water filters, causing 'off flavours' etc. – or even a major economic problem when toxic cyanobacteria may result in a reservoir being unusable as a source of drinking water. However, others see the group as a rich source of valuable chemicals or novel pharmacologically active agents. Approximately 500 species of algae are used as human food or

TABLE 14.1 : BIODIVERSITY OF ALGAE

Algal group	Division	Class	Common name	No.[a]
Cyanobacteria	Cyanophyta	Cyanophyceae	Blue-green algae	2000
Green algae 11000	Charophyta	Charophyceae	Stoneworts	[illegible]
	Chlorophyta	Chlorophyceae Ulvophyceae		3600
	Prasinophyta	Prasinophyceae		120
Chromophyte 000 algae	Bacillariophyta	Bacillariophyceae Fragilariophyceae	Diatoms	1 0
	Chrysophyta	Chrysophyceae Synurophyceae		1000
	Dictyochophyta	Dictyochophyceae Pelagophyceae		10
	Eustigmatophyta	Eustigmatophyceae		12
	Phaeophyta	Phaeophyceae	Brown algae	900
	Prymnesiophyta	Prymnesiophyceae		300
	Raphidophyta	Raphidophyceae		15
	Xanthophyta	Xanthophyceae		600
Red algae	Rhodophyta	Rhodophyceae		4000
Cryptomonads	Cryptophyta	Cryptophyceae		200
Dinoflagellates	Pyrrophyta	Pyrrophyceae		2000
Euglenoids	Euglenophyta	Euglenophyceae		900
Glaucophytes	Glaucophyta	Glaucophyceae		13

[a]Approximate no. of species described.

food products, and about 160 species are considered commercially valuable. Products from algae may have significant financial value; the Japanese harvest of *Porphyra* (Nori) is worth US$1 billion annually and the annual value of algal polysaccharides, primarily agars and carageanans, is US$500 million. Other commercially important products from algae include: health foods ; pigments; aquaculture feeds and lopids. They are also widely used in ecotoxicity testing and may be used to treat waste water. Moreover, algae as a group are responsible for fixing 40 per cent of the earth's carbon and as such are major carbon sinks and also oxygen producers.

ALTERNATIVE STRATEGIES EMPLOYED TO CONSERVE ALGAE

As with other groups of organisms two basic options are available for the long-term conservation of algae: conservation *in situ* in managed or non-managed ecosystems, and *ex situ* conservation. The former has the advantage that the algae will continue to interact with the other biological and physico-chemical factors in their environment and will not vary from 'wild-type strains. This type of algal conservation occurs in marine parks, or other areas protected from the excesses of man's activities. However, in reality this approach is not appropriate for many organisms. Where access to an algal strain is needed quickly, or an axenic culture is required, *ex situ* maintenance is the only realistic option. Further stimuli to the *ex situ* conservation of living materials have been the Convention on Biodiversity (CBD), specifically Article 9: *ex situ* conservation and the parallel development of bioprospecting for products with commercial value.

ROLES OF GENETIC RESOURCE CENTRES AND CULTURE COLLECTIONS

The primary role of an algal culture collection is the same as any other collection of living material that is to be a repository for cultures. In microbial service collections, including algal collections, this role is often associated with other products and services including: provision of authentic specimens for research; material for education; material for bioassay use; aquaculture starter cultures; identification; training; acting as a depository for patent purposes; consultancy; and other commercial applications. All of these require the maintenance of viable, healthy, physiologically and genetically stable cultures.

There are more than 11 000 strains of algae including representatives of approximately 1600 different species retained in protistan collections around the world, with more than 80 per cent of these being maintained in the six largest algal culture collections . These collections provide the scientific community with cultures and their associated information, as well as a variety of other services.

METHODOLOGICAL STRATEGIES EMPLOYED TO CONSERVE ALGAE

Any conservation methodolgy adopted should guarantee long-term stability of the morphological, physiological and genetic characteristics of the preserved organism. A variety of methods have been applied to the long term stabilization/preservation of algae. However, the most commonly used techniques involve the routine serial subculture of the algae under controlled environmental conditions. The alternative approaches, which involve less routine maintenance of the conserved specimens/cultures, all depend on the removal of water and/or altering the cellular physicochemical environment with respect to water activity. These techniques fall into three main categories: drying; freeze-drying and cryopreservation.

Maintenance by Serial Subculture

Historically algae have been maintained *ex situ* by regular serial subculture. This continues to be the method of choice for most phycologists, particularly when relatively small numbers of cultures are involved. There is an extensive literature on culture techniques and maintenance conditions. Medium composition depends both on the requirements of the algae (e.g. diatoms require the inclusion of a silica source) and the preferences of the researcher. Information on medium suitability and full recipes are listed in all major collection catalogues. In general, cultures are maintained under sub-optimal temperature and light regimes (<20°C and <50 μmol photon $m^{-2}s^{-1}$); this maximizes the interval between subcultures and thus minimizes handling/transfers of the strain. Alternative techniques include the use of medium containing organic carbon for maintaining axenic strains capable of heterotrophic or

TABLE 14.2 : ACTIVITIES OF MAJOR ALGAL CULTURE COLLECTIONS

Name of catalogue	Acronym	Country	No. strains	Latest printed	Available on WWW	Use of Cryopreservation	Patent Deposits	WFCC No.
Algensammlung am Institut fur botanik	ASIB	Austria	1600	1985	–	–	–	505
American Type Culture Collection	ATCC	USA	200	1993	+	+	+	1
Culture Collection of Algae and Protozoa	CCAP	UK	1700	1995	+	+	+	140/522
Provasoli–Guillard Center for Culture Collection of Marine Phytoplankton	CCMP	USA	1450	1997	+	+/–	–	2
National Insitute for Environmental Studies Collection	NIES	japan	1000	1997	–	+	–	591
Sammlung von Algenkulturen	SAG	Germany	1630	1994	–	–	–	192
Culture Collection of Algae at the Univ. of Texas at Austin	UTEX	USA	2100	1993	+	+/–	–	606

a Servic collections retaining > 1000 algal strains with the addition of the American Type Culture Collection (ATCC).

\+ Available/currently used.

+/– Under development.

– not available/not used.

mixotrophic growth and solidified, rather than liquid medium may be employed to maximize the period between transfer to fresh medium.

Although serial subculture has proven very successful, with some isolates being maintained for more than 80 years, it is widely recognized as being sub-optimal. It is a labour and consumables intensive process and the continuing increases in costs act as a stimulus to the development of long-term preservation techniques. Furthermore, this technique can potentially lead to selection of a population which may not be representative of the parent culture. In extreme cases this may include changes in important physiological and morphological characteristics, for example, irreversible shrinkage of diatoms, loss of spines in *Micractinium pusillum* and alteration of pigment composition in a number of algae

Maintenance by Storage in Liquid Medium

Some species of algae, particularly those that may form resistánt structures, may be maintained long-term in biphasic medium . This approach has been successfully employed to preserve algal zygotes and cysts for up to 20 years.

Drying Techniques

Some algae are extremely resistant to desiccation and algal cysts/spores may survive prolonged exposure to dry conditions and high temperatures. Examples of this include air dried soil samples containing *Haematococcus pluvialis* aplanospores that can regenerate fresh cultures after 27 years storage and the cyanobacterium *Nostoc commune* that has been revived from herbaria specimens after 107 years storage.

Drying, generally air drying, may be used successfully for a wide range of cyst forming protists and some strains, e.g. the achlorophylous euglenoid *Polytoma*, are commonly transported as dried material on filter paper. Furthermore, storage of cyanobacterial cultures in dried soil, or non-perfumed cat litter, has been used by some researchers to maintain 'back-up' cultures for periods of several years (Parker, personal communication). However, drying has not been widely applied as a method of long-term conservation of algae in major service collections. This is primarily due to the low levels of recovery for some organisms and the relatively short shelf-life of stored material.

More recent research using a controlled drying protocol demonstrated that the method has some potential for a number of green algae. This study employed equipment which vacuum dried the algae in a suspending medium incorporating protective chemicals including skimmed milk, neutral activated charcoal, *meso*-inositol or raffinose. This approach may be utilizable for several algae, but it is unlikely to be satisfactory for the long-term preservation of more fragile organisms and, as yet, no long-term data on viability have been published.

Freeze-drying Techniques

Lyophilization using conventional freeze-drying equipment and protocols employing 20 per cent (w/v) skimmed milk or 12 per cent (w/v) sucrose as protective agents in the suspending medium has been successfully applied to preserve a range of cyanobacteria and eukaryotic microalgae. This technique has been regularly used at the American Type Culture Collection (ATCC) for a number of organisms. However, levels of post-lyophilization viability may be low, 10^{-2} to less than 10^{-7} per cent of the original level being recorded by McGrath *et al.* (1978). On using this approach at the Culture Collection of Algae and Protozoa (CCAP) low levels or no viability was observed post- rehydration of freeze drying eukaryotic microalgae. The highest level of viability observed was 1 per cent for *Chlorella emersonii*; however, no viability was detected after storage for one year. It is worth noting that *C. emersonii* has previously been demonstrated to survive up to two years storage using this technique. In this material viability levels were extremely low, in the range 10^{-4} to less than 10^{-7} per cent of the original level (Day, unpublished data). Freeze-drying cyanobacteria, using the method of kolkowski and Smith (1995) has proven more successful, with survival of both unicellular and filamentous forms. However, where non-axenic isolates were examined on suspension of lyophilized samples in fresh medium, the protective agents stimulate bacterial 'blooms' that had a deleterious effect on the recovery of the preserved cyanobacterium. Although this technique may be employed, low levels of viability, problems associated with non-axenic cultures and the possibility of selection for a tolerant sub-population have dissuaded the major collections from adopting it as a technique to conserve algae.

Cryopreservation Techniques

The general theory and principles of cryopreservation are outlined by Benson. Cryopreservation is the optimal method of long-term storage of algae, where high post-thaw viability can be guaranteed. At ultra-low temperatures (less than –135°C) no further deterioration of stored material can occur and viability levels should be independent of storage duration measured in decades. Therefore, assuming there are no perturbations in the storage regime, long-term stability of the frozen specimens can effectively be guaranteed. As yet there are no published reports on the genetic stability, or otherwise, of cryopreserved algae. However, a selection of the algae originating from different ecological niches, and from different algal Divisions and Classes, were demonstrated to have retained the same levels of post-thaw viability after up to 22 years storage in the CCAP Cryostore. These factors and the significant savings in costs associated with serial subculture have stimulated all the major collections to consider employing cryopreservation, with most of them currently using or developing the technique to preserve a proportion of their holdings.

In comparison to drying or freeze-drying, cryopreservation can result in high levels of post-thaw viability, with levels in excess of 95 per cent for some members of the Chlorococcales. Where levels of viability are high, the possibility of selecting a preservation tolerant sub-population is minimized and also the time required to re-grow a preserved culture to a suitable density for distribution is restricted. At the CCAP, and some of the other major collections, levels of viability in excess of 50 percent, for non-clonal cultures, are required before cultures are retained only in the cryopreserved state. However, other collections including the ATCC do not impose arbitrary minimum levels of post-thaw viability (Nerad, personal communication). Currently at the CCAP approximately 35 per cent of the algal strains lodged in the collections are maintained in a cryopreserved state. At least a further 5 per cent, primarily those strains with less than 50 per cent post-thaw viability, are retained both in a cryopreserved state and by serial subculture.

TABLE 14.3 : BIODIVERSITY OF ALGAL STRAINS MAINTAINED AT CCAP

Algal group	Division	No. strains maintained	
		Cryopreserved	Total
Cyanobacteria	Cyanphyta	152	228
Green algae	Chlorophyta	433	871
	Prasinophyta	5	118
Chromophyte algae	Becillariophyta	8	97
	Chrysophyta	0	20
	Eustigmatophyta	15	24
	Phaeophyta	1	3
	Prymnesiophyta	1	34
	Xanthophyta	43	55
Red algae	Rhodophyta	4	79
Cryptomonads	Cryptophyta	1	62
Dinoflagellates	Pyrrhophyta	1	30
Euglenoids	Euglenophyta	30	120
	Total	694	1741

Two-step Cooling

Most of the freezing protocols that have been developed utilize a two-step system with controlled/semi-controlled cooling from room temperature to an intermediate holding temperature (–30°C being commonly employed). This allows cryo-dehydration of the cells to occur, prior to plunging into liquid nitrogen (–196°C). The frozen material is then stored in either liquid or vapour phase nitrogen in an appropriate liquid nitrogen storage system. Although some organisms can be successfully cryopreserved and stored at higher subzero temperatures (>–70°C), viability levels rapidly fall on storage. Therefore it is necessary to maintain frozen cultures at extremely low temperatures, optimally in liquid nitrogen at – 196°C.

The development of most algal cryopreservation protocols has been empirical. These techniques have been effective in preserving a relatively limited taxonomic range of algae and have largely been restricted to morphologically uncomplicated or small species. A variety of approaches have been employed to improve post-thaw viability; these may be divided into cooling protocol development and freeze-tolerance improvement. Factors that can be varied to reduce, or prevent, damage on freezing and thawing include: type, concentration and duration of exposure to the cryoprotectant; cooling regime employed; storage temperature/ thermal stability of the cryostore; rate of thawing and post-thaw manipulations. Pre- and post-preservation growth conditions of the culture can be altered to increase tolerance to freezing and aid post-thaw recovery; these include: age of culture; light intensity; incubation temperature; osmotic potential of the medium. (Canavate and Lubian, 1995); nutrient limitation) (Ben-Amotzand Gilboa, 1980) and nutritional mode. The majority of effective protocols use late log/early stationary phase cultures; however, the key factor is the 'vigour' of the culture. Cryopreservation of sen escent, stressed or damaged cells will invariably result in lower levels of past-thaw survival compared to the same protocol being applied to a healthy 'vigorous' culture of the same algal strain. Full, step-wise descriptions of protocols are available in the literature.

Encapsulation Dehydration

Encapsulation in alginate gel followed by dehydration in conjunction with the non-penetrating cryoprotectant sucrose has been used to preserve gametophytes of *Laminaria digitalis*. These had survival levels in the range 25–75 per cent depending on age, sex and stress. This approach has also been employed to preserve a range of microalgae and cyanobacteria. The technique was found to be suitable for six of the seven marine algal examined; however, only one freshwater alga, *Chlorella pyrenoidesa,* survived. This species is extremely robust and should survive most standard cryopreservation protocols. An alternative approach employed at the CCAP to preserve *Euglena gracilis* involved encapsulation in calcium alginate and cryopreservation using a standard two- step protocol. This resulted in high levels of post-thaw viability. The mechanism(s) of the protection afforded by encapsulation are not fully understood; however, it appears to: assist in dehydration of the cells; provide support during the freezing process; protect against freeze-fracture; and possibly provide some protection against free-radical mediated injury by functioning as an exongenous antioxidant.

Vitrification

Vitrification using the method of Sakai has been successfully employed to preserve the multicellular alga *Enteromorpha intestinalis*, with 100 per cent of the thalli surviving exposure to the cryoprotectant solution (PVS2) and subsequent immersion in liquid nitrogen. On applying this approach to other algae including *E. gracilis, Vaucheria sessilis* and *Microcystis aeruginosa* no viability following treatment with the vitrification solution was observed. In addition to the toxicity of the vitrification solution, stability of the vitrified material may be a problem during storage or on thawing temperature of up to -110°C maybe observed on removal of cryovials from the bottom of a Cryostore inventory system. At this temperature,damage could theoretically occur to stored material, as it is above the limit of ice crystal growth –139°C. In vitrified samples devitrification and/orice crystal growth could occur at temperatures as low as – 130°C; this could potentially result in cellular damage. In vitrified *E. intestinalis* freeze-fracturing of the thallus occurred during thawing; however, all filament sections were capable of regenerating new thalli. In other systems including mammalian tissues this event has been identified as lethal and it is probable that for unicellular algae freeze-fracture would be lethal.

MECHANISMS OF CELL DAMAGE ASSOCIATED WITH CRYOPRESERVATION

Successful empirically developed cyropreservation protocol are effective on the basis that they reduce osmotic stress, cold shock and potential damage by intracellular and extracellular ice formation, before and during freezing, and on thawing. Improvements to existing methods and the preservation of a greater diversity of algae require a greater understanding of the mechanisms of the fundamental modes of cell damage during freezing and thawing.

Conventional two-step cryopreservation has not proven effective for many of the larger, more morphologically complex, or multicellular algae examined. On employing super-optimal cooling rates, which allow insufficient time for dehydration of the cell, intracellular ice may be formed. In all reported cases in algae, with the exception of *Chlorella prothecoides* intracellular ice formation (IIF) was lethal. By manipulation of the cooling and cryoprotectant regimes it may be possible to avoid IIF, however, in some algae, e.g. *E. gracilis,* even under optimal conditions a proportion of the cells undergo IIF. The filamentous Chlorophyte *Spirogyra grevilleana* was killed by intracellular ice and gas bubbles on being cooled and frozen using a standard protocol at – 10°C min^{-1}. The filamentous diatom *Fragilaria cortonensis* was lethally injured at fast cooling rates by intracellular ice and at slow rates by freeze-induced hypertonic stress. Furthermore gametophytes of *vndaria* pinnatitida which were reported to be successfully cryopreserved only survived four days post-thaw, indicating that a significant amount of cell damage occurred during the process. In the coenocytic alga *Vaucheria sessilis*, on cooling using a conventional cooling protocol: –1°C min^{-1} ⇒ –35°C ⇒ Liquid N_2 [5 per cent (w/v) DMSO], lack of cellular compartmentalization allowed the propagation of intracellular ice throughout the thallus, resulting in death of the alga.

In addition to the above, significant damage has been observed at the ultrastructural level with physical disruption of cellular organelles and membranes. Other factors causing both lethal and sublethal injuries include: pre-cooling manipulations (e.g. centrifugation), cryoprotectant toxicity and chilling damage. These effects can most readily be detected employing vital staining, measurement of oxygen evolution capacity or gross changes in morphology e.g. flagellar loss. Furthermore, free-radical mediated damage and fluctuations in antioxidant levels have also been implicated in freeze-induced damagein in both plant and animal systems. Recent studies indicate that this may be an important factor in the apparent freeze-recalcitrance of some algae.

CONCLUSION

There are limitations in our current understanding of the modes of cryopreservation induced damage and specific sites of injury in algae. Future research will invariably involve the use of techniques including flowcytometry, cryomicroscopy and electron microscopy. In parallel, studies on oxidative stress/injury, other freeze-induced biochemical injuries and the responses of the alga's endogenous protective mechanisms (notably level and composition of antioxidants) to cryopreservation are required.

It is anticipated that elucidation of the key sites of injury will assist in the improvement of existing cryopreservation methodologies and the development of alternative approaches. Areas that present significant challenges include the cryopreservation of large/complex unicellular and multicellular algae. An additional challenge is the improvement of techniques to assess viability. In most published studies, survival has been determined on the basis of post-treatment growth, reaction to vital staining, fluorescence or loss of pigmentation. All of these approaches have shortfalls, regrowth is difficult to assess for non-unicellular algae and other techniques may significantly overestimate post-thaw viability. However, techniques including flowcytometry, measurement of oxygen evolution and response to specific stimuli, e.g. wound healing in *V. sessilis*, may form the basis of alternative rapid viability assays.

In conclusion, although large proportions of the holdings of all the major collections are nominally freeze-recalcitrant, it is probable that if resources were available the majority would be amenable to cryopreservation. The ultimate challenge is to develop approaches that are robust, reliable and result in high levels of post-thaw viability.

15

Bio-Conservation

Within the equatorial region, rain forest covers more than 40 per cent of the land, providing a rich sanctuary for both flora and fauna. Rain forests are priceless reservoirs of plant gemplasm and play a vital role in maintaining global environment stability. The composition, structure and ecology of natural forests, with their large array of plant and animal species is very complex and still remains to be fully understood. Malaysia, as one of the top 12 countries in the megadiversity league, consists of 0.2 per cent of the world's land area but is estimated to harbour more than 2800 species of trees (Saw, 1992) or 6 per cent of the world's flowering plant species. A majority of the rain forest species, especially timber trees, produce recalcitrant seeds. The importance and difficulty of conserving recalcitrant seeded species are well recognized. In recent years, awareness over the loss of plant diversity has captured worldwide attention. Though several conservation activities of plant genetic resources have been performed, they are very labour intensive and complex and encompass various tasks such as germplasm collection, storage and evaluation. Germplasm conservation has become necessary for future sustainable harvesting systems and as a means of maintaining species diversity to prevent genetic erosion. Presently, whilst recalcitrant species can be conserved *in situ* in national parks and forest reserves, some have been taken out of their natural habital and conserved *ex situ* in aboreta and botanic gardens. In the field of biotechnology, many other complementary methods for *ex situ* conservation have been explored, including cryopreservation. According to Ahuja (1991), the success of biotechnological approaches is largely dependent upon three factors:

- The ability of germplasm to survive storage treatments (i.e. dehydration/desiccation, low and ultra low temperature stress).
- *In vitro* morphogenetic competence.
- The development of widely applicable, routine and economically viable conservation methods.

Biotechnology therefore, has the potential for application in crop improvement and genetic resource management programmes. Among other reasons for conservation is the need for providing a continuous supply of seeds and seedlings for the various planting programmes. The current trend the world over is to establish forest plantations in order to reduce the logging impact on natural forests with greater emphasis towards the conservation of the existing biodiversity. In fact, one of the major constraints with tropical forest management is the availability of good quality seeds. For the establishment of seccessful forest plantations there is a need for quality planting materials that are selected from elite mother trees in order to produce seedlings that have improved survival and produced greater economic returns.

SEED CHARACTERISTICS

The majority of tropical tree species have very complex life cycles. Many of these species produce recalcitrant seeds. These species have a long juvenile phase and may come into first flowering and fruiting only after 15-20 years of growth or more. In addition, fruiting and seed sets in many tropical species is not an annual occurrence. When it does occur, the seed production period continues for a short period, which is termed as a mass flowering season that normally occur once in 3–7 years. It is also not possible to predict their flowering times owing to their erratic flowering patterns. Consequently, it causes problems to secure large quantities of fruits on a regular basis. Generally, the seeds that are produced do not undergo dormancy, but instead they are metabolically primed for immediate germination, as soon as the seeds mature on mother plants. According to studies done by Hong and Ellis (1990) and Finch-Savage (1992), desiccation tolerance in recalcitrant seeds increases during seed development on the mother plant; however, unlike orthodox seeds, maturation drying to low moisture contents does not occur. The moisture content range of fresh recalcitrant seeds can be from as low as 36 per cent (*Hevea brasiliensis*) to as high as 90 per cent (*Sechium edule*). Hence, the seeds shed from trees in a fully matured stage undergo rapid post-collection deterioration if the correct environments for continued growth are not provided. The rate of deterioration is dependent on the environment condition around the seeds. This phenomenon leads to reduced rates of germination and seed growth, decreased ability to germinate under stressful conditions and increased probability of abnormal seedlings and longer field emergence. Tropical forest seeds can be broadly divided into three major groups, based on their sensitivity to desiccation and to low temperatures, as follows.

Orthodox Seeds

This group includes all the seeds which desiccate naturally on the mother plants. The seeds can be dried to lower moisture contents (less then 10 per cent) without any serious deleterious effects. In fact, the lower the seed moisture content and storage tepmerature, the longer they survive.

Recalcitrant Seeds

This group of seeds easily die if they are dried below certain moisture limits (12–30 per cent). They also killed when exposed to low temperatures (less than 16°C). Even at optimal moist conditions, survival of seeds in this group is limited from a few days to a few weeks. If they are not collected at maturity and sowed immediately, they will die. The period between collection and sowing is very short. These therefore are difficult to store and do not conform to the rules applicable to orthodox seeds. To date raising planting material from recalcitrant seeds is best done in nureseries but this involves high costs for maintenance and space.

Intermediate Seeds

This is yet another category that has been recently defined. The seeds in this category have storage characters intermediate between orthodox and recalcitrant. Intermediate seeds can be dried to seed moisture levels almost similar to that of orthodox seeds without their viability being affected. However, the dry seeds are easily injured when exposed to low temperatures and viability drops rapidly while in storage. Therefore these seeds can be stored under conditions used for orthodox seeds but only for a short period.

SEED COLLECTION AND HANDLING

For conservation purposes, recalcitrant seeds should be collected from healthy trees with good shape and form. Several techniques of seed collection, namely ground collection, shaking seed bearing branches, free climbing (normally done by natives) or climbing the trees using equipment can be done. Records from Seed Technology Laboratory FRIM show that seeds meant for conservation are best collected freshly from the tree crown by climbing. Criteria for collection of good seeds should include mature seeds, uniformity in colour and size, and healthy. Most recalcitrant seeds respire intensively because of their high moisture content, and hence require good ventilation. If large quantities are closely packed, suffocation, physiological breakdown, fungal growth and overheating will occur, resulting in rapid death of the seeds. On the other hand, the seeds will also deteriorate rapidly if moisture content is reduced too much or too rapidly. This is likely to happen during transporting in open vehicles due to air movements. Plastic bags should be used with their top left open or small holes perforated in the sides. Hessian or jute bags that have been loosely woven are also suitable for transport. Temperatures below 16°C or above 32°C should be avoided for such seeds. Seeds should be kept shaded from direct sun at all times during transport to the conservation centres. Collection journeys should not be more than three days. If it is unavoidable, efforts must be made to dispatch the seeds to their destination by the second day of collection. Any germinated seeds during the collection should be separated from the batch as these can be planted right away at the nursery and later be useful as a source of shoot tips for cryopreservation. It is important to keep the materials moist during transit. At FRIM, a prototype mobile seed–seedling chamber has been invented for the transporting of tropical seeds, especially the recalcitrant type over long journeys, i.e. more than three days with minimal deterioration to the seeds. This vehicle is incorporated with sensors that control the chamber environment, including temperature (20±5°C), humidity (80–95 per cent), and light (if needed).

SEED STORAGE

In the past, many types of storage methods have been proposed for recalcitrant seeded species, but without exception their use has been limited or unsuccessful. Thus to conserve recalcitrant seeds, availability of seeds should be given priority as this will determine the period of storage as viability is reduced during storage.

Short-term and Mid-term Storage Methods

These can be applied to seeds for which viability is maintained for less than 12 months. This duration may be sufficient in cases where short-term conservation is required to 'hold' the germplasm for short periods during which land preparation is required or for the advent of suitable weather for planting. The strategy is to desiccate recalcitrant seeds to just above a critical moisture content, treat them with an effective fungicide and then store them in semi-sealed packaging which allows adequate gaseous exchange, but restricts moisture loss. Several common conventional storage methods include: imbibed storage using sawdust, ground charcoal, pearlite and vermiculite; storage in airtight containers or partial vacuum; regular ventilation or maintaining oxygen levels above 10 per cent; and incorporating germination inhibitors into the storage system. For many other recalcitrant seed species this approach has not shown any significant results. The lowest safe moisture content of many of the recalcitrant species falls within the range of 20–60 per cent.

However, fungicide treatment followed by partial desiccation and storage under ambient or low temperatures have yielded some fruitful results for some crop seeds such as *Hevea* and *Cacao*. Hor (1984) found that after treating *Cacao* seeds with 0.2 per cent Benlate/Thiram mixture, partially desiccating the seeds by air drying and storing them loosely packed in polythene bags at 21–24°C,he was able to prolong the viability of the seeds from one week to about 24 weeks with a 50 per cent germination. Similarly, soaking *Hevea* seeds in 0.3 per cent Benlate, air drying and storing them loosely packed in perforated polythene bags at ambient temperatures, was able to prolony seed viability from three months to one year with about 50 per cent germinability. Although this approach has potential for recalcitrant seed storage, it provides only a short- to medium-term storage strategy.

Long-term Storage

It is evident that with the present limitations, the maximum period of storage that can be achieved using conventional storage methods is about a year or so for recalcitrant seeds. At FRIM, three methods have been devised as protocols for long-term storage: seedling chamber conservation; storage of seedlings on the forest floor; and cryopreservation.

Seedling chamber

Marzalina *et al.* (1994) reported the possibility of using a seedling chamber to slow down the germination and growth of seedlings of recalcitrant seeds. Through this procedure, seed viability in a slow growth condition can be maintained for periods longer than a year. Freshly collected seeds are surface treated with a 0.1 per cent Benlate/Thiram fungicide and allowed to germinate under ambient conditions in containers kept moist using wet tissue papers. Once the radicle has emerged, they are inspected for insect attack or fungal infection. Infection-free germinated seeds are loosely packed in polythene bags or plastic boxes containing sterilized sand or lined moist tissue paper and stored in a specially constructed seedling chamber. This chamber has a temperature humidity and light control and an alarm system. The temperature is maintained at 16°C, the relative humidity at 80 per cent and the photoperiod maintained for four hours each day. The germinated seeds placed in these chambers develop slowly and barely attain the height of 20–25 cm. To date, 20 recalcitrant seeded species have been tested and the storage period obtained was between four and 24 months. When these seedlings are removed and planted in polybags, they require weaning in at least 70 per cent shade for a period of 2–3 weeks before they can be placed under direct sunlight. Post-storage survival percentages are between 60 and 80 per cent when they are planted out. This method of storage is advantageous when there is an excessive seed fall and also it can be used to tied over germplasm during long peroids of unfavourable weather and delayed replanting programmes. However, the disadvantage is that a large facility is required and seedlings have to be hardened before field planting.

Conservation on the forest floor

The other approach being adopted at FRIM is the storage of seedlings on the forest floor under natural subdued light; this method is not new and has been practiced by nature over time. In this approach, areas of the forest floor are cleared and freshly collected seeds are sown in these areas; they then germinate and develop into seedlings. Owing to the subdued light conditions, the seedlings develop very slowly and can remain within manageable heights for up to a few years. Over a period of five years, seedlings of *Hopea odorata* do not grow above 10 cm under these conditions. Once these seedlings are potted into polybags and placed under nursery conditions, they rapidly expand and begin to increase in height. Seventeen species have been tested using this method and the results, to date, are very encouraging. This approach could be adopted to keep stock of planting materials with minimal cost and space. Although this method has good potential, many problems need to be resolved, including protection against field pests and diseases, and improvement of transplanting techniques.

CRYOPRESERVATION

Cryopreservation protocol development has now reached the stage sapplicable for conservation of recalcitrant seeds.

Cryopreservation of whole seeds

Orthodox or semi-recalcitrant seeds of tropical species tested can be cryopreserved with out much problem as their lowest safe moisture content can be reduced to less than 20 per cent. However, when truly recalcitrant seeds are dried and directly plunged into liquid nitrogen, there is no recovery. As mentioned earlier, many tropical recalcitrant seeds are commonly large in size, highly variable in water content, and often attached to appendages adapted for dispersal and thus they will not tolerate a reduction in water content below a relatively high level without loss of viability. Therefore. rapid and systematic approaches are required to initially screen diverse tropical tree seed germplasm for their ability to survive pre-growth, cryoprotectants and dehydration treatments as well as exposure to liquid nitrogen.

TABLE 15.1: SUMMARY OF NON-RECALCITRANT SPECIES THAT HAVE BEEN SUCCESSFULLY CRYOPRESERVED

Species	Part cryopreserved	Temperature/method	MC	Viability
Forest				
Alstonia angustiloba	Whole seed	–20°C, deep freezer	5%	81%
Bambusa arundinacea	Whole seed	–196°C, desiccation	10.7%	73%
Dendrocalamus brandisii	Whole seed	–196°C, desiccation	7.3%	48%
Dendrocalamus membranaceus	Whole seed	–196°C, desiccation	8.5%	53%
Dialium platysepalum	Whole seed	–20°C, deep freezer	10.6%	37%
Dipterocarpus alatus	Whole seed	–196°C, desiccation	10%	68%
Dipterocarpus intricartus	Whole seed	–196°C, desiccation	6–8%	70%
Dyera costulata	Whole seed	–196°C, direct plunge	6.7%	37%
Hopea odorata	Embryonic axis	–40°C, –196°C,	6%	15%
Melia azaderach	Whole seed	–20°C, deep freezer	12.4%	43%
Pterocarpus indicus	Whole seed	–196°C, desiccation	8%	80%
Swietenia macrophylla	whole seed	–196°C, desiccation	5–6%	63%
Swietenia macrophylla desiccation	Embryonic axis	–196°C, encap;	4–5%	63%
Thyrsostachys siamensis	Whole seeds	–196°C, desiccation	7.8%	86%
Fruit tree				
Annona squamosa	Whole seeds	–196°C, direct plunge	7.1%	99%
Averhoa carambola	Whole seeds	–196°C, direct plunge	7%	93%

Baccaurea polyneura	Embryonic axes	–196°C, desiccation	11.3%	25%
Baccaurea polyneura	Whole seed	–196°C, desiccation	5–8%	15%
Carica papaya	Whole seed	–196°C, desiccation	10–11.5%	66%
Citrus aurantifolia	Embryonic axes	–196°C, desiccation	9–11%	100%
Citrus halimii	Embryonic axes	–196°C, desiccation	*16.6%*	*100%*
Citrus mitis	Embryonic axes	–196°C, desiccation	14%	80%
Maikara Zapota	Whole seed	–30°C, desiccation	11%	18%
Manilkara sapota	Whole seeds	–196°C, direct plunge	10.5%	100%
Psidium quajave	Whole seeds	–196°C, direct plunge	10.4%	96%
Acacia mangium	Whole seeds	–196°C,direct plunge	7.6%	76%
Albizia falcataria	Whole seed	–20°C, deep freezer	6.6%	100%
Coffea liberica	Whole seeds	–196°C,desiccation	15–20%	86%
Tectona grandis	Whole seed	–20°C, deep freeser	7–9%	97%
Adenanthera pavonina	Whole seed	–20°C, deep freeser	9.5%	97%
Casia nodosa	Whole seed	–196°C, direct plunge	9.5%	17%
Cassia spectabilis	Whole seed	–196°C, direct plunge	9.5%	20%
Casuarina sumatrana	Whole seed	–196°C, direct plunge	11.2%	23%
Lagestroemia speciosa	Whole seed	–196°C, direct plunge	9.8%	87%
Lagestroemia floribunda	Whole seed	–20°C, deep freezer	12.8%	24%
Livistona chinensis	Embryo	–196°C, desiccation	20%	55%
Prychosperma macarthurii	Embryo	–196°C, desiccation	19%	65%
Veitchia merilli	Embryo	–196°C, desiccation	17%	80%

Cryopreservation of Excised Embryos

Advances in cryopreservation now offer many possibilities for conserving the embryos of recalcitrant seeds. Excised embryos are used for storage because they are small more resistant to desiccation and relatively uniform in size and moisture content. Results indicate that when embryos are subjected to cryopreservation, there is some recovery. Most of the embryos are dissected aseptically and left to partially desiccate in the laminar flow for 4–6 hours, during which time the moisture content can be reduced to 4–33.5 per cent. Krishnapillay and Engelmann (1996) suggested that a more precise and reproducible desiccation could be achieved by placing plant material in a stream of compressed air or in an air-tight container with silica gel. Optimal survival rates are generally noted when embryos are dehydrated down to 10–20 per cent water content (fresh weight basis). The embryos can then be placed in cryovials for storage. Direct immersion of these vials in liquid nitrogen has been practiced but some methods also involve a slow-cooled technique in which the temperature is lowered at a rate of – 1°C per minute to –40°C followed by direct plunging into liguid nitrogen. Techniques for rapid thawing using a 40°C water bath or slow thawing in a laminar flow cabinet do appear to have an effect on the survival of the embryos. The recovery of the embryos can be low (5–15 per cent) using these procedures and viable seedlings grow very slowly. The recovery of the embryos can be culture-media dependent. In some cases modified re-growth patterns such as callusing or incomplete development are observed. Other serious problems include the formation of abnormal seedlings and microbial contamination; the latter appears to be due to seed-borne fungi. Modified recovery conditions, notably the hormonal balance of the culture medium, can significantly improve the survival rate of cryopreserved material.

TABLE 15.2: SUMMARY OF RECALCITRANT SPECIES THAT HAVE BEEN SUCCESSFULLY CRYOPRESERVED

Species	Part cryopreserved	Temperature/method	MC	Viability
Arthocarpus heterophyllus	Embryonic axes	–196°C,	15%	60%
cryoprotectant,				
slow cool				
Calamus manan	Embryonic axes	–196°C, direct plunge	7.8%	60%
Elaeis guineensis	Embryonic axis	–196°C, encap;	10%	60%
desiccation				
Elaeis guineensis	Embryonic axes	–196°C, direct plunge	11.7%	50%
Hevea brasiliensis	Embryonic axes	–196°C, direct plunge	16%	87%
Nephelium lappaceum	Embryonic axes	–40°C, –196°C,	25%	40%
desiccation, slow cool				
Shorea leprosula	Embryonic axes	–40°C, –196°C,	4–5%	12%
Shorea macrophylla	Embryonic axes	–40°C, –196°C,	10%	5%
desiccation, slow cool				
Shorea ovalis	Embryonic axes	–40°C, –196°C	8%	7%
desiccation, slow cool				
Sorea parvifolia	Embryonic axes	–40°C, –196°C,	5–7%	10%
	desiccation, slow cool			

Cryopreservation using encapsulation–dehydration has been successfully applied to many temperate and some tropical species.

Thus, encapsulated zygotic embryos are pre-grown for various periods of time in liquid medium with high sucrose concentrations; they are then partially dehydrated in a laminar flow cabinet or using silica gel and subsequently directly immersed into liquid nitrogen. For recovery, samples are usually placed directly under standard culture conditions. Marzalina (1995) found that dehydrating encapsulated embryo axes of *Swietenia macrophylla* treated with 0.5 M sucrose for two days, followed by eight hours desiccation before cryopreservation gave 63 per cent survival levels. She also found that the survival percentage of encapsulated embryos fell with increasing molarity of sucrose solution and the period of drying. Thus, this technique is largely dependent upon the optimization of per-growth dehydration and desiccation treatments. Once excised, the zygotic embryos of tropical tree seeds undergo phenolic oxidation evidenced by rapid browning or blackening of tissue and surrounding medium. This is also exacerbated by surface sterilization procedures. Studies by Benson *et al.* (1996) indicated that encapsulated embryos of *Hopea weightiana* and *Vatica cinerea* subjected to sucrose dehydration and subsequent air desiccation have made it possible to achieve lower moisture contents which make the embryos more amenable to cryopreservation. Using these techniques, they found that embryos did not suffer the deleterious oxidative phenomenon (tissue browning). Thus, this method offers considerable advantages for the preparation of 'stress-sensitive' germplasm for cryopreservation. Normah and Marzalina (1996) did suggest the use of activated charcoal to be incorporated into the recovery medium in order to absorb the toxic phenolic compounds. As mentioned by Benson *et al.* (1996) increasing importance is being given to the role of stress in recalcitrant seed systems and novel approaches to ameliorating oxidative damage in *in vitro* cultures should also be explored for tropical germplasm.

Cryopreservation of Shoot Tips

Cryopreservation of tropical tree shoot tips or apical meristems normally involves using alginate encapsulation techniques mentioned above. Most studies have been performed on tropical orthodox and intermediate species. Thus, axillary buds of *Morus indica*, adventitious buds of *M. alba*, and axillary buds of *Betula platyphylla* var. *japonica* have been encapsulated in alginate beads and stored for 34–80 days at 4°C and have shown no loss of viability. Maruyama *et al.* (1997) have stored encapsulated shoot tips of *Cedrela odorata, Guazuma crinita* and *Jacaranda mimosaefolia* above freezing temperatures at 12°C, 20°C and 25°C and obtained 80, 90 and 70 per cent viability respectively. At the Seed Technology Section at FRIM, work on encapsulation and desiccation techniques is being attempted on excised embryos of *Shorea leprosula*, but to date no significant results have been obtained.

Vitrification is another technique that can be employed to cryopreserve shoot tips and apices. Such a procedure eliminates the need for controlled freezing and enables cells and meristems to be cryopreserved by direct transfer into liquid nitrogen. To date, most studies have been carried out on temperate species and monocots. As vitrification involves placing samples for pretreatment inextremely concentrated cryoprotective solutions and ultra-rapidly freezing them the intracellular solutes vitrify and form an amorphous glassy structure; the formation of detrimental intracellular ice crystals is avoided. Applications of this technique to shoot tips of apple (*Malus* spp.) and pear (*Pyrus* spp.) from cold-hardened plantlets resulted in 80 per cent and 70 per cent shot tip survival respectively after 40 days after re-plånting. Little work has been reported for tropical recalcitrant species and some are sensitive to the toxic cryoprotectants used in the vitrification mixtures.

Cryopreservation of Somatic Embryos

Both encapsulation–dehydration and vitrification techniques can be employed to cryopreserve somatic embryos. Little work has been performed on tropical recalcitrant seeds (e.g. as an alternative means of conserving tree germplasm) at the moment and this may not be due to the failure of the cryopreservation techniques, but to the problems relating to somatic embryo induction by tissue cultures and the regeneration of these tissues to full plants. Once this can be achieved, no doubt somatic embryos will provide a suitable germplasm source in terms of their response to freezing. They have the advantages of being small in size and being useful for cloning. Micropropagation of tropical trees has been reported to be very difficult as these plants are problematic to culture; thus to date, encapsulated shoot tips offer the most effective means of cryo-conservation.

CONCLUSIONS

The need for cryopreservation becomes evident when one examines conventional preservation systems and their limitation. Reports show that many species decline in viability with time under such storage conditions. The potential use of biotechnology for the conservation of tropical tree germplasm is considerable. However, for many tropical recalcitrant species, there are several prerequisites, which have to be fulfilled before *in vitro* conservation becomes possible. With proper reduction of moisture content and controlled desiccation, seed-recalcitrance will greatly benefit the implementation of conservation strategies. Significant progress can be achieved if more collaborative links between research institutes from tropical and temperate countries working in the same area are established.

16

Cytokines in Agriculture

For most of medical history, therapeutic intervention has relied on exogenous factors. These treatments have run the gamut from magical incantations, to animal-and plant-derived substances, and finally to rational drug design by combinatiorial chemistry. Success in treatment has therefore depended on placebo effect, fortuitous similarities in ligands and receptors (e.g., poppy-derived opiates and the opioid receptors in the brain), or a grasp of biological mechanisms at the molecular level. Paradoxically, this increased understanding of the biology of health and disease, as well as the tools that were developed to gain this understanding, has led to the possibility of utilizing the body's own endogenous factors to treat disease and restore health.

One potential target of this therapy is the group of factors collectively known as *cytokines*. In this chapter we will describe cytokines and their diverse use in human therapeutics, and provide an overview of the multitude of experimental and clinical approaches that have been designed to tap into the body's own healing power. Along the way we will also describe some of the known and potential disadvantages of this exciting new therapy. This review is not meant to be a comprehensive listing of the results of individual studies but rather a broad survey of the types of approaches used.

BIOLOGY OF CYTOKINE

The majority of cytokines (also known by their previous name, lymphokines) are glycoproteins secreted by cells, although membrane-associated forms have also been described. Cytokines are generally not produced constitutively but rather in response to cellular activation. Cytokine production is highly regulated in both paracrine and autocrine fashions at the transcriptional and post-transcriptional levels. In addition, the shrot-half-life of cytokine mRNA suggests selective degradation of the message as a result, evidently, of a particular sequence common to many cytokines and proto-oncogenes. Finally, a variety of mechanisms appear to ensure that cytokines remain compartmentalized. These mechanisms all work to gether tolimit the biological activity of cytokines.

Perhaps the most important biological feature of cytokines is that they are highly pleiotropic and redundant in function. That is to say, most of the cytokines described to date have multiple actions, and these actions often overlap with the actions of other cytokines. In addition, cytokines frequently work via cascading mechanisms that allow them to interact with each other both synergistically and antagonistically. These features are critical from a therapeutic consideration because the ultimate consequences of manipulating any single cytokine must be evaluated in the context of this overall network of factors. Other features of cytokines important from a clinical standpoint are that, in most cases, they act primarily at a local level and that they are rapidly cleared from the circualtion. A notable exception to this rule is interleukin-6 (IL-6), whose importance in regulation of the acute phase response has led to a variety of "chaperone" mechanisms for regulating its systemic bioavailability.

Cytokines interact with specific receptors grouped into the hemopoietin super-family, the tumor necrosis factor (TNF) family, the immunoglobulin superfamily, and the tyrosine kinase family. These receptors are generally membrane-bound molecules, although soluble receptors have also been described. Many cytokine receptors share several similar characteristics, including a subunit structure and an association with signal-transducing elements within the cell. Moreover, several of the receptor subunits are shared between various cytokine receptors, which possibly contributes to cytokine pleiotropy and redundancy.

TABLE 16. 1 : THE CONTINUUM OF BIOINFORMATION

Siganal	Example	Route	Source
Nucleic acids	RNA	Local: intracellular	All living cells, viruses
Cytokines	Interleukine-1	Local:intercellular	Diverse cells
Neuropeptides	β-Endorphin	Local and systemic	Specialized cells
Hormones	Insulin	Systemic	Endocrine glands
Pheromones	Androstenone	Interspecies:external	Exocrine glands
Vocalization	Distress call	Inter/Intraspecies: external	Specialized organs

A frequent conceptual mistake is the belief that cytokines are an exclusive product of the immune system. In fact, certain cytokines are phylogenetically ancient, genetically conserved, and highly pleiotropic molecules (probably the best examples are IL-1 and TNF); forms of these molecules are found in invertebrates lacking a true immune system. Furthermore, IL-1, TNF, and related molecules are intimately involved in the process of apoptosis, a fundamental biological process. Put another way, cytokines should be recognized for their role as conveyers of bio-information rather than as simple effector molecules involved in a single process. illustrates the role of cytokines in the continuum of biological information. This brings us to a crucial detail: cytokines should not, indeed cannot, be seen as operating in a biological vacuum. Rather, as suggested by cytokines may be thought of as a member of a triad including (at a minimum) neuropeptides and hormones and possibly the so-called peptide growth factors as well. It is now well established that these molecules form a complex network responsible for regulating many physiological processes. Certain cytokines (e.g., IL-6) mediate autocrine functions in the nervous and endocrine systems. Likewise, cells of the immune system produce hormones (e.g., prolactin) and neuropeptides (e.g., endorphins); in addition, receptors for hormones and neuropeptides are found on lymphocytes

Thus, given the redundant and pleiotropic nature of cytokines themselves as well as their role as regulatory molecules for various physiological processes, it becomes apparent that developing therapeutics based on modifying the functions of cytokines is far from straightforward. As detailed later in this chapter, toxicity and other unintended consequences are often the result of initial forays into this area. Nevertheless, great advances have been made and continue apace.

TABLE 16.2: THE (ABBREVIATED) UNIVERSE OF HUMAN CYTOKINES

Interleukins (IL)	**Chemokines**	
	α (C-X-C)	β (C-C)
IL-1 α, β	IL-8	RANTES
IL-1RA	gro (α, β, γ)	I-309
IL-2	GCP-2	eotaxin
IL-4 – IL-7	ENA-78	MIP-1 (α, β)
IL-9 – IL-18	SDF-1	MCP (1,2,3)

Colony-stimulating factors (CSF)	**Hematopoietins**
IL-3 (Multi-CSF)	Erythropoietin
G-CSF	Thrombopoietin
M-CSF	Stem cell factor
GM-CSF	flt3 ligand

Tumor necrosis factors (TNF)	**Interferons (IFN)**
Tumor necrosis factors (TNF-α)	Type I (IFN-α, IFN-β)
Lymphotoxin (TNF-β)	Type II (IFN-γ)

Miscellaneous
Oncostatin-M
Leukemia inhibitory factor
Transforming growth factor beta

Table 16.2 illustrates a suggested categorization of the various buman cytokines. Cytokine nomenclature is fraught with confusion; beginning with interleukin- 1, many newly discovered

cytokines were named sequentially regardlesss of their function or genetic relationship to other cytokines. In most current classification schemes, cytokines are grouped according to function or genetic similarity to each other. is not meant to be inclusive; on the contrary, many excellent classification schemes based on function have previously been published.

A brief explanation of the classification scheme used here may be helpful. In general, the term *interleukins* refers to cytokines exerting their primary effect on cells of the immune system; IL-1 and IL-6 are notable exceptions to this broad categorization because they mediate a tremendous number of effects besides immunomodulation. *Chemokines* are small polypeptides that are important not only in immunity and inflammation but in other regulatory processes as well. *Colony stimulatin factors* (CSFs) are important as mediat9ors of hematopoiesis but also exert effects on the immune system. *Hematopoietins* also regulate the function of bone marrow but have fewer effects on the immune response. *Tumor necrosis factors* (TNFs) regulate a panoply of biological effects. *Interferons* were originally named on the basis of their ability to interfere with viral replication, and thus they function as important mediators of host defense; however, they also affect various other processes. Finally, we may lump together **miscellaneous** cytokines that do not fit quite so neatly into the scheme owing to their multiplicity of action. Not included in this scheme are the various *peptide growth factors* such as epidermal growth factor and insulin-like growth factor—whose functions appear to be primarily related to growth, tissue repair, and homeostasis—but that may function as cytokines in certain circumstances. All of these cytokine classes, to varying degrees, represent viable targets for therapeutic manipulation.

CYTOKINES IN HUMAN HEALTH AND DISEASE

Given the ubiquitous role of cytokines in normal human biology, it is only natural that disruption in cytokine levels could be either the cause or effect of human disease. A major advance in understanding the role of cytokines in disease, and particularly in diseases incorporating an immune component in their etiology, was first described by Mosmann et al. (1986). These investigators found that T-helper lymphocytes could be categorized functionally (although not, as yet, phenotypically) into at least two subsets on the basis of their particular pattern of cytokine production. Originally described in vitro using T-cell clones, these subsets were named T-helper-1 and T-helper-2 (subsequent to this, the nomenclature Thl/Th2, T1/T2, and Type 1/Type 2 has also been used; TH1/TH2 will be the preferred usage in this chapter). Building on the data collected from *in vitro* clones, researchers found that these patterns of cytokine production resulted *in vivo* in rodent models. More recently, the TH1/TH2 dichotomy has been confirmed in humans.

- Cytolytic Function
- Help for IgM, IgG, IgA
- Inflammation
- Microbial Defense
- Autoimmunity

- Help for IgG, IgM, IgA, IgE
- Induce CD30 Expression
- SLE
- Atopy

Figure. 16.1

In general, TH1 cells predominantly produce interferon-gamma (IFN-γ), IL-2, and TNF, whereas TH2 cells primarily make (L-4, IL-5, (L-10 and IL-13; these patterns appear to be essentially the same in humans. Interleukin-2 acts as an autocrine growth factor for TH1 and TH2 cells, although TH1 cells are much more sensitive to this activity. Interferon-gamma acts to suppress the growth of TH2 cells. Conversely, IL-4 acts as the autocrine growth factor for TH2 cells, wheras IL-10 suppresses the growth and function of TH1 cells. Various other cytokines are produced in common by these subsets, albeit at relatively lower levels. It should be noted that these distinctions are somewhat fluid, for clones have been isolated in both human and laboratory animal cells that display intermediate cytokine production profiles. Moreover, CD8 (cytotoxic–suppressor) T cells also produce cytokines, as do macrophages and B cells, thus complicating the elucidation of a "typical" cytokine response. However, the general pattern described above for T-helper cells appears to hold true in most cases.

The induction of a TH1-type response versus a TH2-type response (i.e., a state characterized by the preponderance of either TH1- or TH2-type cytokine production) depends on a variety of factors, including the type of antigen-presenting cell, the type and amount of antigen, and factors in the cellular milieu such as other cytokines and hormones. This particular pattern of stimulatory–inhibitory– autocrine functions is now recognized to form the basis for differential immune responses as well as certain pathologies. For example, in normal immune responses TH1 cytokines tend to favor cytolytic (cell-mediated) reactions and the activation of macrophages, whereas TH2 cytokines provide T–cell help for the production of to various antibodies. On the other hand, TH1 responses are usually associated with au–to immune diseases (e.g., multiple sclerosis and rheumatioid arthritis), whereas TH2 responses include atopic conditions, reduced response to infectious challenge, and even sucessful pregnancy. In current and future therapies involving cytokines, it will be of vital importance to understand this interlocking network of cytokines in greater detail to allow possibly for the direct manipulation of immune responses.

Immunomodulation

Ironically, although cytokines are usually associated with the immune system, direct modulation of the immune response has not been the principal therapeutic application of these factors to date. A wide range of options are theoretically available for modulating the immune system, and most of these options are described in this chapter.

Neoplastic Disease

One of the most thoroughly investigated therapeutic uses of cytokines thus far has been in the treatment of neoplastic disease.

Owing to the intrinsically complex nature of neoplastic disease, it follows that the use of cytokines for treatment is equally complex. Therapeutic options to date include disruption of autocrine growth, enhancement of natural immune resistance, the use of cytokines in combination with other drugs, and direct tumor cytotoxicity by cytokines.

It is recognized that many tumors, particularly neoplasias of the hematopoietic system (e.g., leukemia) may involve the autocrine growth of cytokine-responsive or cytokine-producing cells. For example, IL-6 is thought to have the potential to induce polyclonal B-cell activation. In such cases, selective reduction in the levels of certain growth factors could theoretically be beneficial (although clinical improvements). Coversely, deficiencies in cytokine production may impair the function of cellular immune components required for resistance to tumors, such as cytotxic T lymphocytes or natural killer cells. In such cases, the administration of carefully selected cytokines or cytokine combinations may boost a failing immune response.

An intriguing concept that has been explored is the combined administration of cytokines with cytotoxic drugs. Experimental data suggest that administration of cytokines may result in enhanced susceptibility to cytotoxic drugs under certain circumstances. Certain cytokines may also exert a direct cytotoxic effect on tumor cells— either as single agents or in combination. Finally, a relatively successful use of cytokines in oncology has been the repopulation of bone marrow following transplant subsequent to chemotherapy or radiation.

Transplantation

Another successful therapeutic use of cytokines has been as an invaluable adjunct to transplantation. Ironically, successful treatment may take the form of either cytokine suppression or cytokine addition. In the former situation, cytokines are highly active participants in the rejection process responsible for the failure of organ transplantation and may contribute to engraftment failure by a variety of mechanisms. Successful line-term engraftment has been made possible by new drugs that selectively inhibit cytokine production without completely eliminating host defense (discussed in the section on cytokine inhibition therapy).

Conversely, the availability of recombinant cytokines (both natural and modified) has led to great advances in bone marrow transplantation and, more recently, stem cell transplantation. Cytokine therapy may be used in a variety of ways for bone marrow transplantation. For example, administration of colony-stimulating factors and other hematopoietic cytokines (particularly in combination) can be used to enhance repopulation of bone marrow following chemotherapy or radiation treatment and thus lessening potential morbidity from infectious disease or coagulation dysfunction. Alternatively, cytokines may be administered before therapy to enhance the success of autologous transplantation—particularly of stem cells. This expansion of stem cells is usually carried out *in vivo*, although studies are under way to use *ex vivo* expansion of cell populations for transplant.

Infectious Diseases

Resistance to infectious organisms is a primary function of the immune system and therefore requires the precise interplay of most of the cytokine catalog. However, the complex and redundant mechanisms of natural and acquired host resistance are probably not amenable to a simplistic addition or deletion of individual cytokines. Improvements in the control of infectious disease will more likely involve cytokines as a component of the therapeutic regimen.

One promising avenue is the use of cytokines as adjuvants. Although not as powerful as some of the best experimental adjuvants, cytokines have been shown to be at least equal, if not superior, to adjuvants currently approved for human use. Much work remains to be done to optimize this approach; some initial attempts have included using combinations of cytokines, increasing exposure time, and physically associating cytokines with antigens.

Another use of cytokines in treatment and prevention of infectious disease is to correct defects in immune function and thus to enhance host resistance to opportunistic infections common in immunocompromised hosts. This approach usually entails the use of colony-stimulating factors to restore depleted bone marrow, although other cytokines such as IL-1, IL-6, and IFN-* display potential benefits. A more ambitious possibility is the manipulation of cytokine levels (along with other variables) to direct the immune response selectively to either a Type 1 or Type 2 response. At present, this approach remains experimental.

SPECIFIC THERAPEUTIC OPTIONS

Exogenous Cytokine Therapy

"Natural" Cytokines

The discovery of many of the earliest known cytokines took place in human or animal cell cultures, and elucidation of their *in vitro* and *in vivo* effects utilized culture supernatants of stimulated primary lymohoid or myeloid cells. These cultures contain a diversity of cytokines that are produced naturally in response to cellular activation. Although these "natural" (contrasted with man-made) cytokines obviously produce desired biological effects, they have not been pursued as intensively as recombinant cytokines for several reasons. First, mammalian cell culture does not routinely produce cytokines as economically as bacterial fermentations, and mammalian cells are technically more difficult to work with than bacteria or other biotechnology workhorses such as insect cells. However, a more important consideration may be the undefined nature of these cells—particularly the primary cell cultures. These culture supernatants often contain a multitude of cytokines and other bioactive molecules. This undefined nature makes it difficult to determine the activity of individual factors, which is an undesirable circumstance for conventional drug discovery efforts.

On the other hand, a teleological assessment would suggest that just such an assortment of cytokines is perhaps the best application of cytokines and that the cellular source of the cytokine mix represents the ultimate best judge of an appropriate "mix." On the basis of this logic, a mix of natural cytokines may represent an extension of the body's own healing process. One such preparation that has demonstrated clinical efficacy is Leukocyte

Interleukin Inj., which has the trade name Multikine. Multikine is a serum-free lymphokine mixture (containing, among others, IL-2) prepared from human buffy-coat mononuclear cells. Multikine appears to enhance the function of natural killer cells and cytotoxic T lymphocytes. Injection of this product into humans with head and neck cancer was found to result in tumor regression with infiltration of the tumors by lymphocytes. Products such as Multikine may hold promise as adjuncts to lymphokine-activated killer cell therapy.

Recombinant Natural Sequence Cytokines

At present, the overwhelming majority of preclinical and clinical studies of exogenous cytokine therapy have utilized natural sequences produced recombinantly. The literature on the results of these studies is therefore extensive, and a recounting of their results is well beyond the scope of this chapter.

Mutant or Synthetic Cytokines

As described later in this chapter, toxicity is often the limiting factor in protocols involving cytokine admininstration. Seeking to broaden the therapeutic index of these agents, some investigators have found that rational protein drug design can result in cytokines with enhanced biological activity and reduced "changes in quantitative parameters." This latter phrase was proposed by Brouckaert *et al.* (1994) as an alternative to "toxic side effects" in describing the sequelae of exogenous TNF administration and implies a belief that the current state of knowledge does not allow for a precise understanding of wheter observed affects are truly toxic or simply adaptive changes. This semantic device is undoubtedly cold comfort to patients enduring the side effects of cytokine therapy (see the section on toxicity associated with cytokine therapy).

The majority of published work on cytokine mutant proteins (muteins) has focused on synthetic versions of TNF. Tumor necrosis factor is a potent, pleiotropic cytokine; because it is involved in a multitude of both normal and pathological processs, it represents an obvious target for cytokine therapy. However, this very diversity of actions has proven to be problematic when TNF is administered systemically, which often results in profound toxicity. This has led to efforts to design TNF molecules that retain beneficial activity but minimize the toxic side effects. Several investigators have reported success in this effort, although clinical data are not yet available.

Another recent example of this approach is the construction of synthetic cytokine (Synthokine) SC-55494, a potent IL-3 receptor antagonist first described by Thomas *et al.* (1995). By employing oligonucleotide-directed mutagenesis on a synthetic, natural sequence IL-3 DNA followed by iterative combination, these investigators produced several thousand IL-3 mutants that were subsequently screened for biological activity. The results of this screening identified SC-55494 (Synthokine), a molecule with 48 amino acid sequence changes relative to natural IL-3. Synthokine was found to produce a ten- to twentyfold increase in hematopoietic activity, although its potential for induction of inflammatory responses was only about twofold greater than natural sequence IL-3. Subsequent primate studies deomonstrated that administration of Synthokine, either alone or in combination with G-CSE, singificantly enhanced recovery from both neutropenia and thrombocytopenia following radiation-induced bone marrow apla-sia. These results suggest that Synthokine and related molecules may hold singnificant promise in the support of myelosuppressed patients.

Several other cytokine muteins have been constructed and are under evaluation for the treatment of various conditions. Examples include IL-4 muteins and IL-5 mutein for the treatment of allergic diseases.

Fusion proteins

A different form of cytokine is the fusion protein. The concept behind fusion proteins is to connect the functional portion of a cytokine (i.e., the receptor-binding moiety) with another molecule; this second molecule can have biological activity of its own or may be there to assist the function of the cytokine molecule. In the case of fusion proteins consisting of two different cytokines, the resulting molecule is termed a hybrid cytokine or hybrikine. One of the first examples of a hybrikine was PIXY321 (Pixykine), which consists of the active domains of IL-3 and GM-CSF coupled by a flexible amino acid linker sequence. By incorporating elements of both IL and GM-CSF (both of which are potent colony-stimulating factors), PIXY321 was found to stimulate wide range of progenitor cells. Perhaps more significantly, PIXY321 have shown it to be a useful adjunct in autologous bone marrow transplantation among other indications. Other cytokine–cytokine fusion proteins are now being described such as CH925, which is a fusion protein comprising rhIL-6 and rhIL-2. This hybrid molecule exhibits erythropoietin-like activity.

Cytokines may also be combined with other molecules. For example, construction of cytokine–antibody hybrids produces fusion proteins retaining the properties of both parents (i.e., antigen binding and cellular activation). Such constructs display promise as potent immunostimulants or as treatments for conditions such as septic shock. A different but related immunostimulant property can by obtained by fusing cytokines directly to antigens, as demonstrated by kim *et al* (1991).

An interesting example of a novel problem that was overcome by constructing a mutant cytokine was reported by Anderson et al. (1997). These investigators worked with IL-12, a cytokine that has been somewhat problematic to produce recombinantly owing to its heterodimeric structure and because the distinct subunits of the molecule are produced by separate genes on different chromosomes. This problem has been overcome by constructing a single-chain fusion protein (Flexi-12) that retains the biological activity of the natural cytokine molecule. No doubt this approach will find greater utility in the future.

Methods for Delivering Cytokines

As previously mentioned, cytokines are proteins or small polypeptide molecules; therefore, oral administration has been impractical because these molecules cannot withstand the digestive environment. Their administration thus far has been primarily parenteral — mainly by intravenous injection. Although certainly the most convenient method clinically, this approach is associated with many serious disadvantages, including the necessity of bolus administration (or at least short-term infusion), which induces toxicities, as well as the need to administer high doses of cytokines to yield sufficent locally effective concentrations (i.e., lack of targeting).

A variety of advanced methodologies are under development for a more efficient and clinically manageable delivery system for therapeutic cytokines. Although many of these techniques are in the developmental phase, they show exciting promise as future treatment options. illustrates some of the methods now under development in this area.

Cytokine gene transfer

One type of gene transfer is designed specifically for treating neoplastic disease. In this approach, cytokine genes are inserted directly into tumor cells, which subsequently produce cytokines. Cytokines exhibit several properties that make them good candidates for treatment of neoplasia, including immunostimulatory activity, the ability to enhance tumor antigen presentation or antigenicity, and direct cytotoxic activity. However, the toxicity associated with systemic cytokine treatment (discussed later in this chapter) is a major drawback. On the other hand, a more localized administration might lessen this toxicity.

In practice, tumor cells from a patient would be removed, specific cytokine genes would be inserted, and these cells would then be reinfused. The theory behind this technique is that the tumor cells will secrete cytokines *in vivo*; when these cells are encountered by immune effector cells, the secreted cytokines will enhance the antitumor effect, and the tumor cells will be eliminated. Although some success has been demonstrated in experimental animal systems, some cytokines actually appear to enhance tumor growth rather than suppress it. Another potential problem is that gene transfer may induce growth autonomy in the tumor cells and actually exacerbate the problem. Clearly, many issues remain to be addressed before implemention this technique in human therapy.

Cytokine Inhibition Therapy

Chemical (Nonbiological) agents inhibiting cytokine production and action

A variety of nonbiological drugs have been found to suppress cytokine production either specifically or nonspecifically. These drugs act via myriad mechanisms, including alteration in cytokine gene transcription or mRNA translation, inhibition of cytokine release, or inhibition of cytokine processsing. To date, the primary clinical use of these agents has been as adjuncts to organ and tissue transplantation, in which they have provided great advances in clinical success. Increasingly, these agents are now being evaluated for treatment of other immunerelated diseases. Some of these agents are discussed in the following paragraphs.

Glucocorticoids are well known and powerful immunosuppressive drugs and enjoy wide clinical use for a variety of indications in which inappropriate immune reactions are factors in disease etiology. These drugs act at a variety of molecular targets; germane to the present discussion is their role in modulating cytokine production. Glucocorticoids affect cytokine production via a number of mechanisms including transcriptional repression, posttranscriptional alterations, induction of the suppressive cytokine TGF-β, antagonisms of transcription factors, and cooperation with transcription factors. Glucocorticoids also alter the expression of cytokine receptors. Owing to their highly potent anti-inflammatory and immunosuppressive effects, these drugs are associated with numerous side effects and toxicity.

Cyclosporin (SandImmune) is probably the most thoroughly studied of the immunosuppressive drugs that specifically target cytokine production. Cyclosporin is a fungal metabolite with several unique features including a novel amino acid in its composition. Cyclosporin acts primarily at the level of the T lymphocyte and effectively and reversibly inhibits the action of these cells. The drug exerts this highly specific effect by binding to the cytosolic protein cyclophilin, which is a member of a class of molecules generically known as immunophilins. The cyclosporin–cyclophilin complex targets calcineurin and essentially binds calcineurin to the immunophilin. The calcineurin is unable to ineract with downstream transcription factors such as NF-AT, which ultimately prevents the transcription of cytokine genes, including IL-2 IL3, IL-4, IL-5, and TNF.

Tacrolimus (FK506) is another immunosuppressive fungal metabolite. Like cyclosporin, FK506 seems to exert its principal effects by altering the expression of cytokine genes—in particular the gene for Ll-2. This agent shares little structural similarity with cyclosporin, although it has a similar mechanism of action. Like cyclosporin, FK506 acts by binding to a cytosolic immunophilin;unlike cyclosporin, for FK506 this immunophilin is termed FK506 binding protein (FKBP). Inhibition of IL-2 gene transcription then occurs in a molecular manner similar to that following treatment with cyclosporin.

Rapamycin (Rapamune, Sirolimus) shares structural similarities, including identical binding domains, with FK506; like FK506, it also binds to the cyclophilin FKBP. However, unlike its molecular cousin, rapamycin appears to have only limited (or no) effect on cytokine production but rather appears to affect cell cycle progression in late G1 phase by inhibiting growth factor signal transduction pathways. Although rapamycin does not appear to affect cytokine production, there is some evidence that it may decrease the stability of cytokine RNA within the cell. Work on elucidating the precise mechanism of this drug is under way.

Leflunomide (HWA 486) is a synthetic immunomodulatory drug (an isoxazole derivative) that has been demonstrated to be effective in animal models of autoimmunity and transplantation rejection. Early reports on its mechanism of action were conflicting, although it was recognized at the outset that the drug operates via different mechanism than the macrolides described above. In particular, early studies provided conflicting data regarding its ability to suppress cytokine production or the expression of cytokine receptors. More recently, Cao *et al.* (1996) have demonstrated that leflunomide does in fact work via modulation of cytokine function and thus enhances the production of the immunosuppresive cytokine TGF-β. It is clear that more work is required to identify the drug's exact mechanism of action.

Thalidomide has languished for many years owing to its teratogenic effects following its use in pregnant humans. In recent years, however, this drug has demonstrated potential as a useful inhibitor of cytokine production and thus may be enjoying a comeback. For several years thalidomide has been known to be a potent suppressor of TNF production and has recently been demonstrated to suppress production of the immunoregulatory cytokine IL-12. The mechanism of cytokine suppression by thalidomide is currently unknown, although it may act as an immunomodulator (but not necessarily an immunosuppressant) via selective gene regulation. Thalidomide is currently being evaluated for clinical treatment of several conditions such as autoimmune disease and graft-versus-host (GVH) disease.

Pentoxifylline is a methylxanthine drug that has reproducibly been shown to inhibit the production of TNF and IL-12 and to affect the production of other cytokines such as IL-1, IL-6 IL-8, and IL-10 variably. Pentoxifylline is showing some promise in maintenance of AIDS patients, although several other clinical conditions may be candidates for treatment with this drug.

Pentamidine is an antiprotozoal drug used primarily to treat pneumonia caused by *Pneumocystis carinii* as well as certain/inflammatory conditions. In addition to its antiprotozoal activity, pentamidine has been demonstrated to inhibit the production of IL-1 via a posttranslational event— possibly altering the cleavage of the precursor form of the cytokine. More recently, pentamidine has been shown to inhibit the production of inflammatory chemokines.

Tenidap is an antirheumatic drug that combines cyclooxygenase inhibition with suppression of the acute phase response. Tenidap has been demonstrated to suppress the production of IL-1 IL-6, and IFN-γ. It appears specifically to suppress cytokine production by altering ionic homeostasis, although the exact mechanism of action remains unknown.

Other drugs have also been shown to inhibit cytokine production. The precise mechanism of action of many of these drugs is currently unknown, and they have not been developed extensively for clinical use as specific cytokine inhibitors. Some of thes agents are listed in. As knowledge of the molecular biology of cellular activation grows, drugs will undoubtedly be formulated to modulate specific mechanisms.

TABLE 16.3 EXPERIMENTAL DRUGS WITH ANTICYTOKINE ACTIVITY

Drug	Reference
Phophodiesterase isozyme inhibitors	Yoshimura et al. 1997
Metalloproteinase inhibitors	Gallea-Robache et al. 1997
p38 kinase inhibitors	Badger et al. 1996
SK&F86002	Triplett et al. 1996
MDL 201, 449A	Edwards et al. 1996
CGP 47969A	Casini-Raggi et al. 1995

Biological Agents that Inhibit Cytokine Production or Action

Specific inhibitory cytokines

To date, only one naturally occurring cytokine antagonist has been identified, namely interleukin-1 receptor antagonist (IL-1RA), which exists in two forms: secreted and intracellular. This antagonist is a member of the IL-1 cytokine family (which includes IL-1α and IL-β) according to amino acid sequence, receptor binding avidity, and gene structure and location . This agent seems to function as a pure receptor antagonist and does not exhibit any discernible agonist activity (e.g., internalization of receptor complexes, etc). Experimental evidence suggests that IL-1RA plays a role in various disease states such as rheumatoid arthritis (RA), sepsis, diabetes, and other diseases. Understandably, the clinical application of a natural receptor antagonist represents an ideal clinical tool; clinical trials are under way for treatment of RA.

Mutant proteins have also been constructed that act as a specific receptor antagonist for both IL-4 and IL-13 activity. This early success should lead to the creation of additional specific cytokine antagonists that are threrapeutic.

Cytokine receptors

Although most cytokines appear to exert their biological activity via interaction with specific cell-surface-bound receptors, many cytokine receptors are known to be present in a soluble form in the circulation of normal, healthy individuals. These soluble receptors may bind their cognate antigen with the same affinity as the surface receptors and have been postulate to have various physiological functions. One possible function is to serve as a chaperone molecule protecting the cytokines as they are ferried through the circulation. Another possible function, and one that serves as the basis for their possible use as therapeutics, is as a binding molecule for free cytokines in the circulation, which serves to control their bioavailability. Most of the work in this area has involved the IL-2 receptor, although several other soluble receptors have been investigated as well.

A potential problem (or benefit, depending on the desired outcome) inherent in the use of soluble cytokine receptors is their propensity for actually enhancing the activity of certain cytokines— possibly by extending their presence in the circulation. Pharmacokinetic studies with recombinant cytokine receptors have demonstrated variability in the clearance between the different receptors as well as with different forms of the same receptor. Given the structural and functional difference between the different cytokine receptor subfamilies, this is perhaps not very surprising. However, this finding does point to the need to evaluate these molecules on a case-by-case basis in their development as human therapeutics.

Anticytokine antibodies

Like soluble cytokine receptors, naturally occurring anticytokine antibodies have been demonstrated in the circulation of normal healthy individuals. Also, like the soluble receptors, these antibodies are thought to represent a mechanism for controlling the bioactivity of cytokines and to prevent them from inducing inappropriate responses. These molecules have been invaluable research reagents in understanding the role of cytokines in myriad biological processes and are increasingly being investigated as therapeutics.

Several anticytokine antibodies (particularly anti-IL-4 and anti-IL-5 antibodies) have been evaluated with some success in the treatment of allergic disease. A variety of other conditions have been treated with anticytokine antibodies with varying degrees of success. As with the addition of exogenous cytokines, treatment of disease with single anticytokines antibodies may be confounded by the highly redundant nature of cytokines.

Fusion toxins, immunotoxins, and chimera toxins

In a slightly different use of anticytokine biotherapeutics, the cytokine-producing cell rather than the cytokine itself is targeted. As discussed elsewhere, the overproduction of cytokines, or at least an overly vigorous response by cytokine receptor-bearing cells (e.g., lymphocytes), can lead to a number of pathological conditions such as autoimmunity or neoplasia. The greatest degree of specificity could be achieved by developing a so-called "magic bullet," a molecule that is both highly specific and toxic. Potential candidates for this role are fusion toxins, which are also termed immunotoxins or chimera toxins.

Numerous highly potent toxins are found in nature, and three of these in particular have been used in the construction of immunotoxins: diphtheria toxin and *Pseudomonas* exotoxin (both bacterial products) and ricin (a plant product). These molecules share several structural features, including a binding domain allowing them to attach to cells, a component that allows the molecule to cross the target cell membrane, and an enzymatically active domain (the toxophore) that poisons the cell. Unfortunately, the binding domain allows the toxins to attach to many different cell types. However, by removing this binding domain and replacing it with a portion of the cytokine or growth factor to interset the toxicity of the resulting molecule may be tailored to specific targets. The resulting hybrid molecule is now specific only for cells bearing a particular receptor; the molecular components mediating translocation and intoxication remain functional. Two of the earliest successes with this approach were the toxins DAB_{486}-IL-2 and DAB_{389}- IL-2, which contain fragmentary portions of the IL-2 molecule .Clinical trials and experimental studies with these immunotoxins have found them to be relatively safe and effective for a wide variety of conditions, including inhibition of HIV-1 RNA. mycosis fungoides, rheumatoid arthritis, and IL-2 receptor-expressing malignancies.

Several other fussion toxins have been constructed. Most of these novel toxins are still experimental, and this their actual utility as therapeutics remains to be established

TABLE 16.4 VARIOUS CYTOKINE FUSION TOXINS

Toxin	Cytokine	Reference
Diphtheria toxin	G-CSF	Chadwick et al. 1993a
	IL-6	Chadwick et al. 1993b
	IL-15	van der Spek et al. 1995
	GM-CSF	Bendel et al. 1997
	IL-3	Liger et al. 1997
	IL-1	Frankel et at. 1996
	GM-CSF	Burbage et al. 1997
Pseudomonas exotoxin	IL-4	Ogata et al. 1989
	IL-6	Siegall et al. 1988

Antisense technology

With a more complete understanding of molecular genetics has come the ability to alter fundamental biological processes experimentlly and therapeutically. Nowhere is this manifested more than in the emerging field of antisense technology. Antisense technology is based on the construction of synthetic ologonucleotide sequences that will bind specifically with "sense" (i.e., message sequence) nucleic acids and in effect mask them and prevent sequence-specific actions such as translation of mRNA into protein. In theory, antisense therapy could represent a highly specific tool able to interrupt undesired biological processes at their most fundamental level. The advantages relative to cytokine therapy are obvious, for antisense technology allows a highly selective shutdown of specific cytokines or related growth factors to occur while theoretically leaving other factors unaffected. This would make antisense more desirable than even fusion toxins because it involve deactivation—rather than destruction—of a cell.

Although the concept of antisense technology has been demonstrated repeatedly in the laboratory, many important problems remain to be solved before (or if) this technique becomes a routine therapeutic option. These include knowledge of the proper target sequence, target accessibility, specificity, and drug stability and delivery. One potential anticytokine use that appears to have bypassed this last problem is the administration of antisense oligonucleotides topically as an adjunct to wound healing. Another and more serious problem with antisense therapy has been unanticipated toxicity—particularly cardiovascular changes. At present, antisense technology appears to be more valuable as a research tool for understanding cytokine biology than as a component of cytokine therapy.

Extracorporeal removal of cytokines

In situations where in situ inactivation of cytokines may be impractical or impossible, a more brute force option may be the physical removal of cytokines. One approach that has been investigated is removal of cytokines during continuous hemofiltration. To date, the results of this approach have been equivocal; although removal of certain proinflammatory cytokines from the filtrate has been demonstrated, there has been little evidence of a reduction in systemic cytokine levels. Furthermore, improvements in patient outcome do not appear to be clinical trials of this approach are under way.

A more selective approach for physical removal of cytokines from the circulation is by extracorporeal binding to antibodies. This technique was reported by Weber and Falkenhagen (1996) and was termed the Microspheres Based Detoxification System. In essence, polyclonal antibodies specific for IL-1, IL-6, and TNF were covalently linked to microspheres, and human plasma spiked with these cytokines was processed through the system. The authors reported efficient removal of these cytokines using this technique, for the rates of removal were greater than those reported following ultrafiltration. This technique has at least two advantages over removal by cytokines rather than bulk removal of molecules. Second, antibodies to other molecules can also be attached concomitant with the anticytokine antibodies; this might facilitate the removal of other molecules associated with a particular disease state such as endotoxin during sepsis. The clinical effectiveness of this strategy has yet to be demonstrated

MONITORING THE CYTOKINE THERAPY:

With the capability of altering the balance of cytokines in vivo, the accurate determination of the level of these molecules naturally becomes of great importance. Cytokines are routinely measured by one method or a combination of three methods: bioassay, immunoassay, and molecular assay. Although each of these assay types has its own spectrum of advantages and disadvantages, an inclusive comparison is beyond the scope of this review. Nonetheless, be far the most commonly employed type is the enzyme-linked immunosorbent assay (ELISA), which is fairly sensitive (and with certain modifications, highly sensitive), highly specific, relatively inexpensive to perform, and techncically straightforward. Also, ELISAs are easily adapted to automation, which is an obvious advantage in the clinical laboratory. Given their current and almost certain future acceptance, it is prudent to take a closer look at the potential problems and technical issues associated with the use of ELISAs for quantiating cytokine levels in clinical samples. Examples of some of the more critical considerations are as follows:

- Effect of sample processing: Several studies have demonstrated that variables in sample collection, processing, and storage can affect the results of cytokine assays.
- Existence of alternative molecular forms of the antigen: Certain cytokines can exist in variant forms (e.g., IL-6) or as precursor molecules (e.g., IL-1b). The type of antibody used in the ELISA becomes an issue in these circumstances in that monoclonal antibodies (often used in commercial ELISAs) may not necessarily detect the presence of such alternative forms.

- Presence of interfering substances: A wide variety of substances may be present in clinical samples and can interfere with immunoassays. Some of these include soluble receptors, natural antagonists such as IL-1RA, chaper-one molecules such as macroglobulin, and cytokine-specific autoantibodies.
- Assay precision: Although excellent reagents and kits are increasingly available, assay precision is still an important consideration in cytokine measurement. At present the most reliable method for increasing precision is strict attention to methodological details.
- Reference standards: A variable that is receiving increased attention is the source of the reference material used in the assay. Because experiments are ultimately extrapolated back to the reference preparation, this represents a weak link in the assay. However, until recently there has been limited effort to standardize the reference among commerical suppliers of cytokine ELISAs; thus, results obtained with any given kit may not match those obtained with other kits. It is to be hoped that this essential item of quality control will soon be addressed.

Unlike many other biological entities (e.g., hormones, enzymes, etc.) in which normal human clinical ranges have been established, the expected or "normal" values for cytokines in various biological samples have not yet been published. A comprehensive review of the literature suggests that the vast majority of such evaluations have been made to compare cytokine levels present in humans during particular disease states with normal (i.e., appropriately matched control) individuals. A secondary source of such information is the product inserts for various commercial immunoassays kits that quote the results of internal studies performed by the supplying company. Unfortunately, an exhaustive compilation of all available data is still lacking, although such a further compilation will be of great value.

Even if some day technical assays are perfected and comprehensive baseline values are compiled, a final consideration must be made; namely. Which cytokines should be evaluated and under what circumstances? For example, monitoring the levels of an exogenously administered cytokine for pharmacokinetics would be relatively straightforward—particularly if a novel cytokine were administered (e.g., mutant cytokines with a unique protein structure), and might be useful for maintaining "therapeutic levels" (if in fact such levels have been estabilshed).

On the other hand, because baseline cytokine levels are low or undetectable, measuring these factors in the circulation would be essentially useless if one were evaluating potential immunosuppression(either as a disease sequela or clinically induced). In such cases, measurement of ex vivo cytokine production by stimulated cells would be required.

A final consideration concerns the evaluation of cytokine-related biomarkers. For example, neopterin, a biologically stable molecule induced by interferon-gamma , might be used to monitor therapy with this molecule. Conversely, certain cytokines such as α_2-macroglobulin are known to interact with cytokines. Although these molecules do not appear to affect the function or bioavailability of cytokines significantly, more information is needed to assess their contribution fully.

TOXICITY ASSOCIATED WITH CYTOKINE THERAPY

Preclinical toxicology is an integral and vital component of all pharmaceutical and biotechnological drug development. In toxicological evaluation, the potential of a drug candidate to affect various organ systems adversely at up to superphysiological doses, is estimated to establish a margin of safety before the initiation of clinical trials in humans.

With the advent of genetically engineered proteins in recent decades, an era of more "natural" forms of therapy was anticipated. From the early (and, in retrospect, perhaps a bit naive assumption that "if a little is good, then more is better" view of cytokines and other biological therapeutics, a realization has emerged that the road to rational biotherapy will be somewhat more difficult. At the risk of fatal oversimplification, the preclinical and clinical evaluation of cytokines comes down to two points: (1) traditional animal toxicology models may not reproducibly predict the ultimate human toxicity of recombinant molecules, particularly cytokines, and (2) the sequelae of in vivo cytokine administration (and perhaps cytokine suppression as well) can not necessarily be predicated on the basis of their in vitro actions. Let us examine these points in greater detail.

The first problem namely the inability of routine animal models to predict the toxicity of cytokines, can best be understood by considering the basic biology of the cytokines themselves. Three essential features are of particular importance:

1. Certain cytokines (e.g., IL-3, IL-12, GM-CSF, and IFN-g) exhibit a relatively high degree of species specificity. Unlike most nonbiological therapeutics, cytokines mediate their normal physiological (and apparently at least some of their deleterious) effects via their interaction with highly specific receptor molecules. This receptor interaction is also the basis for the strict species specificity of certain cytokines. Thus, administration of such molecules to an inappropriate species may be expected to result in difficulties in interpretation.
2. Although many recombinant proteins are highly homologous to their endogenous counterparts, minor biochemical discrepancies (e.g., base pair substitution, alternative forms of glycosylation, etc.) may result in altered biological function or, more likely, antigenic stimulation and lead to the induction of neutralizing antibodies. Although not as common when cytokines are evaluated in a homologous species (e.g., mouse–mouse), when human sequence cytokines are injected into laboratory animals, neutralization of the protein's bioactivity will result within a few weeks. Consequently, effects that may not be apparent in humans except after chronic administration will effects that may not be masked in animal models.
3. Whereas structure and function have been relatively well conserved for most mammalian cytokines, certain of these

molecules display divergent function between species. In such cases, the clinical sequelae of human cytokine therapy may not be predictable even from appropriately designed animal studies.

In spite of these potential difficulties, progress is being made in preclinical assessment of cytokine toxicity. For example, transgenic animals have been created that have a high constitutive level of cytokine production and thus mimic the effect of super-physiological doses of cytokines in humans. Another approach is the construction of transgenic animals expressing human cytokine receptors.

The second problem, namely that the clinical response to in vivo cytokine administration is not necessarily predictable by in vitro data, is likewise a complex issue. Let us once again examine this in the light of basic cytokine biology:

1. As stressed earlier, cytokines are pleiotropic and redundant in function. An example of this is provided by tumor necrosis factor (TNF), which exhibits potent cytotoxicity for isolated (i.e., *in vitro*) tumor cells and induces the regression of solid tumors following *in situ* injection. However, TNF's broad diversity of physiological actions (not the least of which is the mediation of systemic shock) precludes its use as a sytemic therapeutic. Cytokine redundancy may account for certain toxicities by inappropriately enhancing (or suppressing) the effects of other endogenous cytokines.
2. Cytokines appear to function principally as carriers of bioinformation at a local level. Although cytokines are now known to mediate the functions of the immune, nervous, and endocrine systems (the neuroimmunoendocrine axis), most of these interactions probably take place at discrete sites. This impression is supported by the demonstration that the systemic (circulating) levels of almost all cytokines are extremely low and that exogenously administered cytokines are rapidly cleared from the circulation.

A common unintended consequence of cytokine treatment although not a toxicity per se—is the development of anticytokine antibodies. This may seem surprising intuitively, for cytokines are endogenous molecules. However, anticytokine antibodies have been demonstrated in the sera of normal healthy individuals and are now thought to constitute part of a complex regulatory system for controlling the activities of endogenous cytokines. Thus, it should not be surprising that administration of superphysiological concentrations of these proteins induces the production of specific antibodies. Moreover, as mentioned elsewhere regarding animal studies, even recombinant proteins may not duplicate the natural product, which results in the creation of novel antigens and the subsequent development of antibodies.

Anticytokine antibodies may be either binding (i.e., directed against epitopes with out functional consequences) or neutralizing (eliminating the biological activity of the molecule). Naturally, of the two, neutralizing antibodies represent the greatest concern because their presence could effectively negate the therapeutic benefit of treatment. Owing to the number of studies evaluating these molecules, most of the neutralizing antibody information has been gained from clinical trials of interferons. However, these are not isolated incidents, and minimizing the development of anticytokine antibodies will be an ongoing challenge to clinicians.

Aside from the issue of neutralizing antibody formation, administration of almost all exogenous cytokines tested to date has been associated with a range of toxicities of various severity; some representative examples of such toxicity are listed in Table.This table lists some of the more common toxicities associated with three different classes of therapeutic cytokines; the list of symptoms is representative of such adverse side effects and is by no means a comprehensive catalog.

One toxic side effect that appears to be common to a number of different cytokines (particularly IL-2 and GM-CSF) is the so-called vascular leak or capillary leak syndrome. This syndrome is characterized by such symptoms as peripheral and pulmonary edema, hypoxia, occasional ascites, perhaps hypotension, and sometimes respiratory failure. Although the syndrome is suspected to be of immunological origin, the exact mechanism of toxicity remains to be elucidated.

TABLE 6. REPRESENTATIVE TOXICITIES ASSOCIATED WITH THERAPEUTIC CYTOKINE ADMINISTRATION

Cytokine(s)	Reported toxicities	Reference
Interleukin-2	Hypotension, weight gains, capillary leak syndrome, renal dysfunction, gastrointestinal disturbances, neurological symptoms	Bruton and Koeller 1994
Hematopoietic growth factors	Bone pain, autoimmune disturbances, vascular leak syndrome, arteral thromboses, cutaneous reactions	Vial and Descotes 1995
Interferons	"Flu-like syndrome," neurological toxicity (fatigue, behavioural changes), hypo/hypertension, myalgia automimmunity (rare), tachycardeia, hematotoxicity	Vial and Descotes 1994

CONCLUSIONS

The concept of harnessing the body's own natural healing powers has long been considered the ideal form of medicine. The traditional avenues available for accomplishing this have taken the form of herbs (whether crude concoctions or semisynthetic formulations) and spiritualism (ranging from magical incantations to prayer). However, in recent years, the discipline of molecular biology has provided researchers and clinicians with the highly precise tools to make this promise a reality as last. Genetic engineering allows use to produce products in essentially unlimited amounts, thus making it possible to replace missing biochemicals where a deficiency exists. Like wise, technologies such as antisense and hybridoma-produced monoclonal antibodies allow us to deplete "unwanted" biomolecules selectively.

Nevertheless, as pointed out in this chapter, a comprehensive understanding of the design of the vertebrate organism still eludes man. The myriad interconnected and interdependent biological processes work together in health; it may be the height of naivete to assume that the simple addition or deletion of a single molecule will always effect a cure, diseases such as diabetes not withstanding. As is being reaffirmed in clinical studies, such as the case with cytokines.

Not that success is completely elusive. For selected applications such as bone marrow reconstruction, organ and tissue transplant, and several immunological disorders, cytokines are proving to be an exciting addition to the medical armamentarium. Moreover, certain innovations have resulted in "better than nature" molecules that can be tailored to specific pathologies, and the future promises even more improvements. It is only reasonable to temper our enthusiasm with the knowledge that much remains to be learned.

Index